ROBOTICS:
A PROJECT-BASED
APPROACH

LAKSHMI PRAYAGA, CHANDRA PRAYAGA,
ALEX WHITESIDE, AND RAMAKRISHNA SURI

Cengage Learning PTR

D1126959

CENGAGE
Learning·

Professional • Technical • Reference

Australia • Brazil • Japan • Korea • Mexico • Singapore • Spain • United Kingdom • United States

CENGAGE Learning

Professional • Technical • Reference

Robotics: A Project-Based Approach
Lakshmi Prayaga, Chandra Prayaga,
Alex Whiteside, and Ramakrishna Suri

Publisher and General Manager,
Cengage Learning PTR: Stacy L. Hiquet

Associate Director of Marketing:
Sarah Panella

Manager of Editorial Services:
Heather Talbot

Senior Product Manager: Emi Smith

Project Editor: Dan Foster

Technical Reviewer: Jeremy Branchcomb

Interior Layout Tech: MPS Limited

Cover Designer: Mike Tanamachi

Indexer: Valerie Haynes Perry

Proofreader: Sam Garvey

For product information and technology assistance, contact us at
Cengage Learning Customer & Sales Support, 1-800-354-9706.

For permission to use material from this text or product, submit all requests online at **cengage.com/permissions**.

Further permissions questions can be emailed to
permissionrequest@cengage.com.

Oracle and Java are registered trademarks of Oracle and/or its affiliates.

All other trademarks are the property of their respective owners.

All images © Cengage Learning unless otherwise noted.

Library of Congress Control Number: 2014945698

ISBN-13: 978-1-305-27102-9

ISBN-10: 1-305-27102-5

Cengage Learning PTR

20 Channel Center Street

Boston, MA 02210

USA

Cengage Learning is a leading provider of customized learning solutions with office locations around the globe, including Singapore, the United Kingdom, Australia, Mexico, Brazil, and Japan. Locate your local office at: **international.cengage.com/region**.

Cengage Learning products are represented in Canada by Nelson Education, Ltd.

For your lifelong learning solutions, visit **cengageptr.com**.

Visit our corporate website at **cengage.com**.

Printed in the United States of America
1 2 3 4 5 6 7 16 15 14

About the Authors

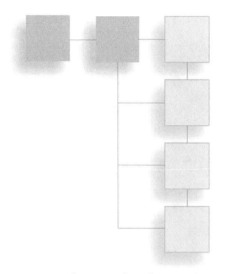

Dr. Lakshmi Prayaga is an Associate Professor at the University of West Florida, Pensacola, Florida. Her research interests include the use of advanced technologies in education, including serious games, robotics, and mobile app development. She has authored and co-authored several articles in international conferences and journals. She has also received several grants to build and implement educational environments using advanced technologies.

Dr. Chandra Prayaga is a professor and chair of the physics department at the University of West Florida. His research interests include study of the properties of liquid crystals, laser spectroscopy, and physics education, particularly in the use of technology such as robotics for teaching physics. He has received grants to train schoolteachers in physics, physical science, and mathematics.

Mr. Alex Whiteside is a software engineer and entrepreneur. Previously employed at American Express Technologies and the U.S. Air Force Research Lab, he now works as a researcher for the University of West Florida and serves as Chief Technology Officer of RILE Inc. He specializes in the development of low-level server applications and user experience design (UX) for web, desktop, and mobile applications.

Dr. Ramakrishna Suri retired as a professor in the aerospace department of the Indian Institute of Science, Bangalore, India. His research specializations include electronic instrumentation payloads for rockets. He has trained several PhDs in instrumentation.

Contents

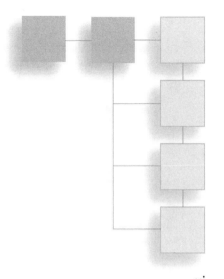

Introduction

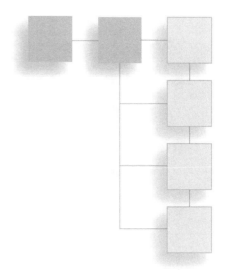

This book is designed to provide an introduction to applications of robotics. It is designed for absolute beginners, such as teens and those who wish to venture into the field of robotics. Several fun-filled activities are presented in the book, including a robot sweeper, a medical assistant, and a security robot.

How This Book Is Organized

This book is organized as a project-based approach to learning robotics. Each chapter includes a project that you can build and complete. Materials required for each project are listed in the beginning of each chapter, and a complete list of all materials for all the projects is provided in Appendix A.

Companion Website Downloads

The source code for all projects is available on the companion website at:
www.cengageptr.com/downloads

A set of videos demonstrating how to build each project is also available for purchase.

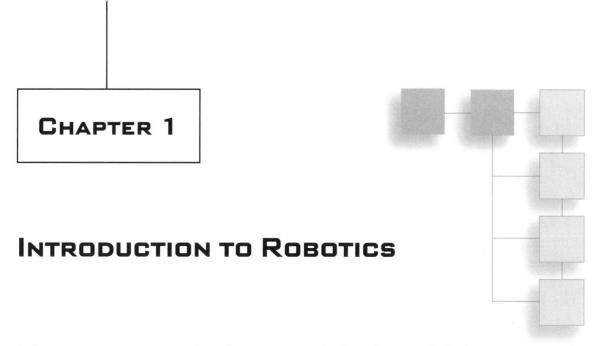

CHAPTER 1

INTRODUCTION TO ROBOTICS

Robotics is an exciting interdisciplinary topic—the foundations of which rest on principles from various disciplines, including computer science, physics, engineering, and mathematics. Applications of robotics are varied and their scope is limited only by human imagination. This chapter presents a discussion of the definition of a robot, a variety of robotics applications, and a quick synopsis of the robotics projects that you will build in each chapter in this book.

HISTORY OF ROBOTICS

The notion of robotics can be traced back to as early as 320 B.C. when the Greek philosopher Aristotle writes in his famous book *Politics*:

> *"If every tool, when ordered, or even of its own accord, could do the work that befits it... then there would be no need either of apprentices for the master workers or of slaves for the lords."*

Since then researchers have been working in the field of robotics for a long time. In fact, the following link provides a neat timeline on the evolution of robotics from researchers at the University of Auckland: http://robotics.ece.auckland.ac.nz/index.php?option=com_content&task=view&id=31

As you can see, the idea of a robot was in use in the 1400s, and the notion of an *artificial being* was introduced in the 1700s. In the 1900s, the word "robot" was first used in the context of a play, and the word robot was described as something that lacks emotions. In the 1940s, British scientists designed the first autonomous machine, and in 1950 came the famous Turing test, in which Alan Turing proposed a test to determine whether machines could think. The 1970s and 1980s produced several advances in the design of robots with the introduction of mechanically controlled arms and other mobile robots that could be used in industry. The 1990s produced human-like robots that could be used in games such as soccer. The new millennium (2000 until the present time) brought in advanced robotics in commercial and household applications. We see robots used as vacuum cleaners, mail delivery agents, surgeons in hospitals, and in many more situations. We also see new designs of robots such as flying drones that could be more suitable for some applications. In fact, robotics is becoming so pervasive in our daily lives that researchers in Auckland, New Zealand, predict that by 2020 the demand for robotics will become so high that there will be a dearth of engineers and programmers to meet this new demand.

WHAT IS A ROBOT?

A quick search on the Internet provides several attempts at defining what a robot is. For our purposes, we define a robot as *a mechanical device that can perform a given task depending on the instructions it is given.* So, how is this different from a computer or a machine? Unlike a computer or a machine that performs a given task, typically, a robot not only performs the task given to it, but it is also able to use artificial intelligence (AI) and *learn* from its experiences while performing given tasks. However, in this book we will limit our discussion to the application of robotics in various disciplines and the understanding of how to design robots that can perform tasks related to specific disciplines or fields such as health care, military, law enforcement, etc.

Robots also come in various shapes and styles designed to accomplish specific tasks. Some robots are shaped like mechanical arms that perform tasks requiring arm-like limbs, such as lifting objects, placing items, drawing on a board, and other such tasks. Some robots are shaped like vehicles, such as unmanned cars, flying drones, underwater ships, and so on. So we see that robots have various shapes and styles for different purposes.

The chapters in this book are aligned to various applications and domains of robotics and include a project that demonstrates how robots are being used in those domains. In this first chapter, you will find a discussion of some applications of robotics in various aspects of real life.

ROBOTS IN COMMERCIAL APPLICATIONS

Robots are making their way into our daily lives. There are several commercial robots suitable for a variety of needs. Chapter 2, "Build Your Own Robot Sweeper," describes how to build a simple version of a commercial application of robotics. A robot vacuum cleaner (such as the Roomba) is an example of a commercial application of robotics, sold as a consumer electronic device. While designing such an efficient robotic vacuum cleaner is a complex process involving sophisticated artificial intelligence algorithms and electronics, it is possible to mimic its basic behavior and design a robotic sweeper using simple components. This chapter guides you in building a simple robotic sweeper, as seen in Figure 1.1.

Dust cloth attachment

Figure 1.1
Robot sweeper.

BASIC ROBOT NAVIGATION

Navigation is one of the basic features for a robot. Chapter 3, "Traveling Robot," introduces basic navigation mechanisms that help the robot navigate using line tracking (Figure 1.2). You will also learn how the robot can detect colors along its path. This feature can become a good option for making the robot take different paths depending on the color detected. You will learn how to use a line-tracking algorithm and a color sensor.

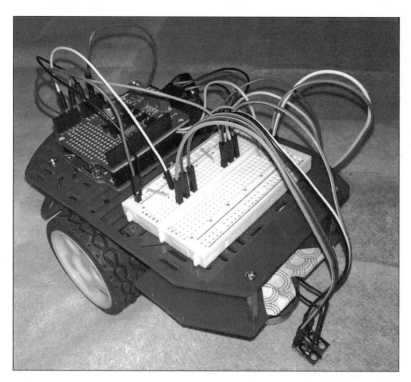

Figure 1.2
Navigating robot with sensors.

Robots in the Military

The role of robotics in the military is extremely important, and robots are being used in ingenious ways in this area. An example of this is the use of drones. In all these applications, one aspect of robotics is of paramount importance: the notion of intrusion detection. This capability triggers the robot to take the necessary defensive actions and thwart danger. Chapter 4, "Intruder Alarm," guides you through building a robot equipped with proximity sensors that can detect intrusion and trigger the robot to raise an alarm. Typical combat robots are equipped with several sensors, including optical, proximity, and infrared sensors, which are able to detect intrusions and raise alarms. Figure 1.3 shows the Arduino microcontroller with the alarm detection sensor attached to it.

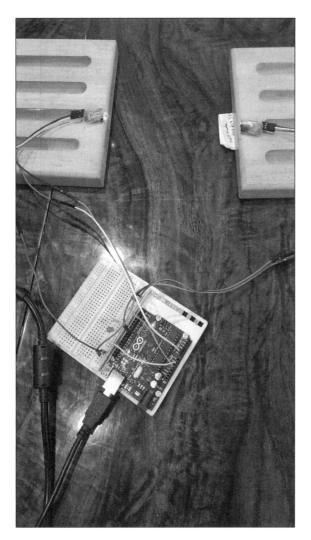

Figure 1.3
Alarm detection sensors.

WI-FI NETWORKING

A principle feature of robots is their mobility, which requires autonomy and un-tethered connections for power, communications, and sensors. While various technologies can provide the wireless features required for modern robots, none match the simplicity and effectiveness of Wi-Fi for basic uses (Figure 1.4). In Chapter 5, "Robot Networking and Communications with Wi-Fi," you will learn about the client–server paradigm, with an emphasis on networked robots. You will set up a basic server on your robot and communicate with it over your home Wi-Fi network to issue commands from your computer, no wires required. This project in Chapter 5 extends all other projects so far, allowing those projects to take place remotely, with data sent to the user's computer without a serial connection.

Figure 1.4
Robot with Wi-Fi shield.

Robots in Medicine

Robots are being used extensively in various capacities in the medical field. Robots are being used both at a very simplistic level as simple data-capturing kiosks in tele-medicine, as well as performing highly skillful surgeries in a physical hospital. In Chapter 6, "Robot Medical Assistant," the assistant project presents you with step-by-step instructions for designing a robot that can be used to remind a patient to take pills at a particular time—essentially, a pill reminder (Figure 1.5). The logic can then be extended to program the robot to monitor the patient's temperature or sugar levels, etc., and truly transform the robot into a personal medical assistant.

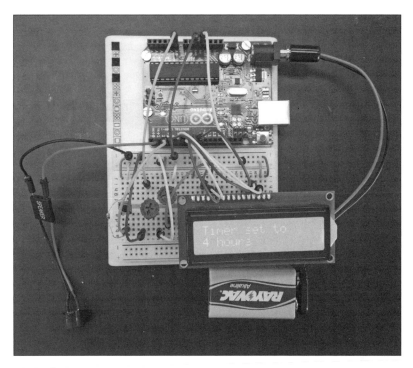

Figure 1.5
Robot with timer capabilities.

WEATHER MONITOR

Weather monitoring is a very complex and highly computationally intensive task. Computers can be very efficient in calculating, analyzing, and making forecasts from given data sets. In weather related problems, meteorologists work with large data sets, model them, and predict weather patterns as we see in cases of hurricanes, tornados, and other weather related issues. Chapter 7, "Data Logger," walks you through building a simple temperature monitoring application where you can monitor the temperature and log it to an SD card (Figure 1.6).

Figure 1.6
Robot with temperature sensor and data logger.

USER INTERFACES

In the modern technical age, text based data has evolved into a much prettier picture. Nearly every consumer application provides a graphical user interface (GUI) to make the solution as easy and fun to use as possible. Chapter 8, "Remote-Controlled User Interfaces," analyzes the leading graphical user interface packages and their use in industry. You will then create a JavaFX application that will connect to your robot over Wi-Fi. You will learn to design beautiful interfaces using FXML- and CSS-based technologies, and then implement the functionality in high-level Java code (Figure 1.7). This code will link to your Wi-Fi server developed for your robot and execute commands and return data at the push of a digital button.

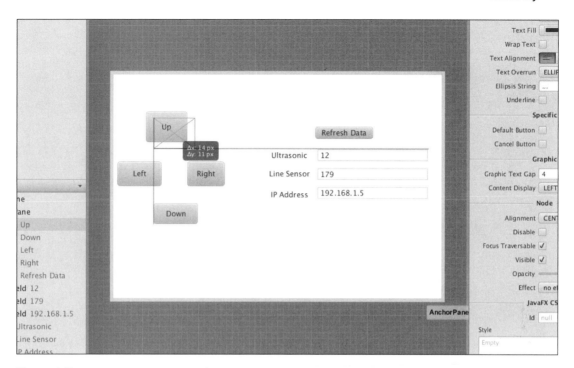

Figure 1.7
UI for the robot.
Source: Oracle.

SECURITY

One of the most remarkable increases in the technology revolution is that of camera sensors. Shrunken down to centimeter-wide lenses with incredible data speeds, just about every device comes equipped with a camera, resulting in a remarkable number of images on a global scale. In Chapter 9, "Security Robot," the project introduces you to a basic camera technology through the development of a security robot (Figure 1.8). This robot is capable of driving to a remote location via the previously created interface and then snapping a picture and sending it over the Wi-Fi network to the GUI, where it will be displayed.

Figure 1.8
Camera equipped robot.

Entertainment

While robots are primarily functional, they also add entertainment value. Chapter 10 will use sensors and lights to create a light and sound system in which multicolored lights dance to music using a NeoPixel ring and a microphone.

Mobile Connections

Previously, all robotics activities required a computer present to send commands or retrieve data. In today's world, mobile phones have created an environment free of physically tethered devices. Chapter 11 explores the development of a mobile app on the Android OS, which can remotely control your robot!

Conclusion

A discussion on current research and applications in the field of robotics is presented in Chapter 12, "Additional Robotics Applications." You will also be presented with possible extensions to projects discussed in this book to encourage further exploration into the field of robotics.

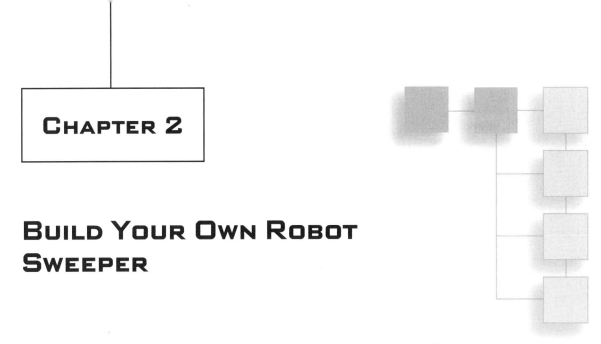

CHAPTER 2

BUILD YOUR OWN ROBOT SWEEPER

In this chapter, you will accomplish the following objectives and build a robot sweeper.

CHAPTER OBJECTIVES

- Build a robot chassis
- Add the Arduino board
- Add the Ardumoto board
- Add the ultrasonic sensor
- Add a piece of dust cloth
- Complete assembly of the robot
- Insert the required code

INTRODUCTION

This chapter describes how to build a simple version of a commercial application of robotics. The Roomba robot vacuum cleaner is an example of a commercial application of robotics, sold as a consumer electronic device. While designing such an efficient robotic vacuum cleaner is a complex process involving sophisticated artificial intelligence algorithms and electronics, it is possible to mimic the basic behavior and design a robotic sweeper using simple components. This chapter guides you in building a simple robotic sweeper.

There are three parts to the activities described in this chapter:

1. Assemble a basic robot with a motor control shield and an ultrasonic sensor, which navigates by avoiding obstacles and works as a robotic sweeper.

2. Program the components of the software that control the motors and the ultrasonic sensor.

3. Put it all together.

MATERIALS REQUIRED

- Arduino Uno R3 board (Amazon, SparkFun, Adafruit)
- Magician chassis kit (Amazon, SparkFun)
- Ardumoto board (Amazon, SparkFun)
- Ultrasonic range sensor Model HC-SR04 (Amazon)
- 9-volt battery
- Jumper wires with connectors
- Solderless breadboard, plug-in type (Amazon)
- Piece of dust cloth (e.g., Swiffer)

PART 1: ASSEMBLING THE ROBOT

You need a chassis, fitted with wheels, which can be made to rotate with the help of motors. You should be able to control the motors with a software program.

Assembling the Chassis

You can build your own chassis with aluminum angle and nuts and bolts from the nearest hardware store. How big should it be? How large should the wheels be? What kind of motors? How much torque? There are, of course, many questions. If you are a do-it-yourself expert, you can go to the nearest hobby shop and start talking to the staff about ideas. Or, you can simply do what we did and buy the packaged components of a robot chassis. Here again, as you begin looking, you will find a wide and bewildering variety. We found that the "Magician Chassis," available from several vendors, is a convenient chassis for most of the robotics projects described in this book. Abundant literature on the Internet describes the chassis, including videos on assembly and several robotics projects that can be done with it. You can purchase the Magician Chassis in kit form and assemble it. Figure 2.1 shows all the parts that come in the kit.

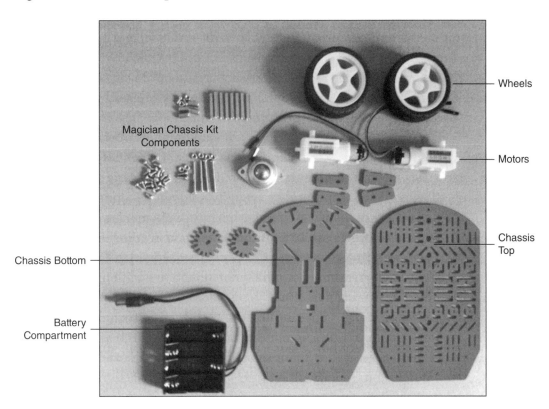

Figure 2.1
Parts of the Magician chassis.

Follow the instructions provided with the kit and assemble the chassis.

Mounting the Arduino Board

After assembling the kit, the next thing to add to the chassis is the Arduino board. Again, there are several versions of the Arduino board. We are using the Arduino Uno R3. You can easily screw the board onto the top of the chassis, or even attach it to the chassis with rubber bands. After completing this step, your robot should look similar to Figure 2.2.

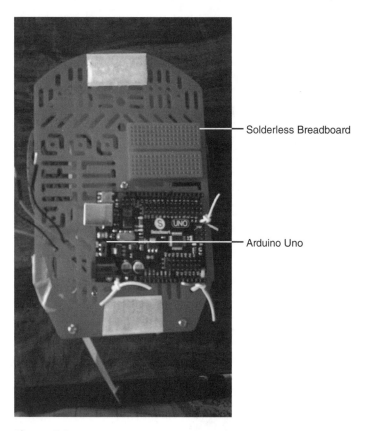

Solderless Breadboard

Arduino Uno

Figure 2.2
Arduino Uno board and solderless breadboard.

The power for the Arduino board can come from a USB cable connected to the USB port on the Arduino board, which is used for communications with your laptop. But once the robot starts moving, you will need a power source onboard the chassis. This can come from a single 9-volt battery. Unfortunately, the usual batteries that we use for flashlights and such around the house do not last long once we connect the motors. We recommend using long-lasting Ni-MH rechargeable batteries such as those available from Tenergy. The battery can also be attached to the chassis with a small strip of Velcro. For convenient connection to the Arduino board, we suggest using a battery clip with a 5.5 mm/2.1 mm plug that plugs directly into the power jack on the Arduino board (see Figure 2.3).

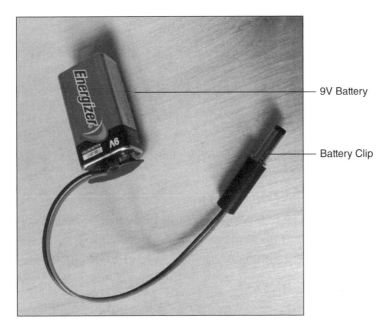

9V Battery

Battery Clip

Figure 2.3
Battery plug.

Mounting the Ardumoto Motor Driver Shield

The Arduino Uno board is the brains of your robot. However, depending on your application, you will need different "shields," which are additional boards that provide additional functionality to your robot. For example, if you want to add Wi-Fi capability to your robot, you can buy a Wi-Fi shield, or, as in the present case, you will need a motor driver shield, so that the robot can, as the name suggests, drive motors.

We used the Ardumoto Motor Driver Shield, available from SparkFun, and found it very easy to install and program. The SparkFun website has an excellent tutorial on how to mount and use the shield:

https://learn.sparkfun.com/tutorials/ardumoto-shield-hookup-guide/assembly-tips

Follow the instructions there and mount the Ardumoto shield. This will require a bit of soldering.

Mounting the Ultrasonic Sensor

You will find that the ultrasonic sensor, like most sensors, is easy to understand and use, so we will mount that and make it work before starting the more complex job of making the robot move.

First, here is a brief description of how an ultrasonic sensor works. This sensor sends out an ultrasonic pulse and receives the pulse reflected from an object in front, such as a wall. The time between the sent and received pulses, multiplied by the speed of sound, gives the distance between the sensor and the wall. We will use this sensor so that our robot can sense when it is too close to the wall and either turn away or stop.

A picture of the sensor is shown in Figure 2.4. There are four pins on the sensor, labeled Vcc, Gnd, Trig, and Echo. These pins should be connected to specific pins on the Ardumoto board, as listed in the Table 2.1. The Vcc pin should be connected to the 5V pin on the board, the Gnd pin should be connected to a Gnd pin, and the Trig and Echo pins should be connected to any of the digital pins except digital pins 3, 11, 12, or 13.

Ultrasonic Sensor

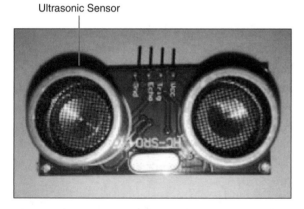

Figure 2.4
Ultrasonic sensor.

Table 2.1 Sensor Pin Connections to Ardumoto Board	
Sensor Pin	**Ardumoto Board Pin**
Vcc	5V
Gnd	Gnd
Trig	Any digital pin (e.g., 7)
Echo	Any digital pin (e.g., 6)

The connections can easily be done using jumper wires. It is convenient to attach a solder-less breadboard on the chassis. The sensor pins can be inserted directly into the bread-board, which also acts as a convenient mount for the sensor, and the jumper wires can all be connected using the breadboard. Here is a helpful YouTube video on how to use a breadboard, in case you are not familiar with it:

http://www.youtube.com/watch?v=zCX3Hr4ZsDg

Figure 2.5 shows the robot with the ultrasonic sensor. You can also see the Ardumoto board mounted on top of the Arduino board.

Figure 2.5
Robot with ultrasonic sensor and Ardumoto and Arduino boards.

Inspect the Ardumoto board that you just mounted and familiarize yourself with the pins along its sides. You should be able to identify the six analog pins numbered 0 through 5, the Ground (GND), the Reset (RST), the 3.3 V, the 5 V, and the VIN pins along one edge, and the digital pins along the opposite edge. Note that the digital pins 3, 11, 12, 13 are labeled PWMA, PWMB, DIRA and DIRB. These are explicitly for motor control and should not be used for any other purpose. For example, you should not connect an ultra-sonic sensor's Trig or Echo pin to any of these pins.

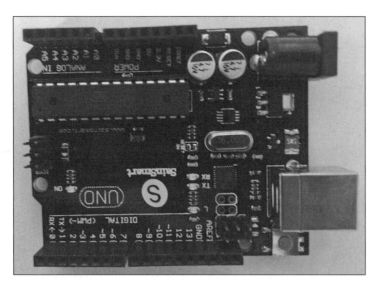

Figure 2.6
Ardumoto board.

After mounting the Ardumoto board, connect the motors to the board. Notice the A 1, 2 and B 3, 4 terminals on the Ardumoto board. These are near the USB connector on the Arduino. Connect the 1, 2 terminals labeled A to the two wires leading to one motor, which from now on we will call MOTOR_A. Similarly, the 3, 4 terminals labeled B are connected to the other motor, henceforth to be called MOTOR_B. It does not matter whether you connect the red wire to 1 and the black wire to 2 or the other way around for either motor. The way you make the connections will however define the direction the motor turns: clockwise (CW) or counter-clockwise (CCW). You can always interchange the red and black wire connections later, if necessary.

This completes the assembly of the robot, which should now look like the one in Figure 2.7.

Wires to the motor

Ardumoto board

Battery clip

Figure 2.7
Completed robot.

PART 2: THE SOFTWARE

In this section, you will begin to write the software required for your robot to perform the tasks assigned to it.

Using the Arduino IDE

You will use the Integrated Development Environment (IDE) available with Arduino to write, test, and save all your programs. Details of how to set up the environment and use it are available at the following URLs.

Instructions for setting up the IDE to use Arduino for Windows are available here:

http://arduino.cc/en/guide/windows

Instructions for setting up the IDE to use Arduino for Mac OSX are available here:

http://arduino.cc/en/guide/macOSX

You will need to visit the appropriate URL and set up the IDE so you can start using it below.

We will learn how to control the motors using the Ardumoto motor control shield in small steps following the instructions given below.

There are two ways of completing this project. One is to just take our complete code, load it into the robot, and have fun watching it run around. The other is to experiment with each part of the code so you will learn what each part does, and then put it all together yourself. This way, if you want to design your own project, you will know exactly what to do. For example, you might want to add a couple of extra motors to control a claw that can pick up objects on the floor and drop them in the trashcan. You might also want to mount your ultrasonic sensor on a rotating motor so that it can sense objects all around the robot in addition to objects in front of it, effectively giving the robot the ability to "turn its head" and "look" around.

We certainly suggest the latter method, and the instructions for this approach are given below.

Controlling the Speed and Direction of a Motor

Reacquaint yourself with the motor control pins on the Ardumoto board. Pins 3 and 11 control the power to motors A and B. The power is determined by a number, of type `byte`, which can take any value from 0 to 255. The command

```
analogWrite(3, 150);
```

makes MOTOR_A run with a power corresponding to the number 150. The command

```
analogWrite(11, 150);
```

makes MOTOR_B run with the same power.

Pins 12 and 13 control the direction of rotation for motors A and B. The direction is determined simply by a digital value (0 or 1) given to the appropriate pin. The command

```
digitalWrite(12, 0)
```

puts the value of 0 on pin number 12, and makes MOTOR_A turn clockwise (CW). The command

```
digitalWrite(13, 1)
```

places the value of 1 on pin number 13, and makes MOTOR_B turn counterclockwise (CCW).

Code to Control the Motor

Comments within the code are preceded by two forward slashes, //, at the beginning of the line, as in the following three lines. These comment lines are for our explanations; the program just ignores these lines.

// Clockwise and counter-clockwise definitions. Instead of having to remember which number,
// 0 or 1, corresponds to which direction, we define the symbols CW to mean 0 and CCW to mean 1.
// Depending on how you wired your motors, you may need to swap red and black.

Open the IDE for Arduino and type the code given below. Save it with the filename singleMotorControl.

```
#define CW 0
#define CCW 1

// Motor definitions to make it easier to remember:
#define MOTOR_A 0
#define MOTOR_B 1

// Pin Assignments. Instead of remembering pin numbers, we give descriptive names to them
// Don't change these pin numbers! These pins are statically defined by the shield layout of
// the Ardumoto.
const byte PWMA = 3;   // PWM control (speed) for motor A
const byte PWMB = 11;  // PWM control (speed) for motor B
const byte DIRA = 12;  // Direction control for motor A
const byte DIRB = 13;  // Direction control for motor B

void setup()
{
  setupArdumoto(); // Set all pins as outputs and initialize them to LOW
}

void loop()
{
  delay(2000);
  digitalWrite(DIRB, CW);
  analogWrite(PWMB, 50);
  delay(2000);
  analogWrite(PWMB, 0);
}
// setupArdumoto initializes all pins
void setupArdumoto()
{
  // All pins should be set up as outputs:
  pinMode(PWMA, OUTPUT);
  pinMode(PWMB, OUTPUT);
  pinMode(DIRA, OUTPUT);
  pinMode(DIRB, OUTPUT);
```

```
// Initialize all pins as low:
digitalWrite(PWMA, LOW);
digitalWrite(PWMB, LOW);
digitalWrite(DIRA, LOW);
digitalWrite(DIRB, LOW);
}
```

How to Use the Code

Upload the code into a sketch and then play with the parameters in the `loop()` function. Change the speed, the direction, and the motor, and watch as each motor obeys your commands. Make sure that CW means the same direction of rotation for both motors. If necessary, you can interchange the black and red wire connections to one of the motors to achieve this.

Running Both Motors Using Function Calls

The following code segments are for illustrative purposes. You will find all this code included in Part 3, "Putting It All Together," which contains the complete code including the functions discussed in this section.

Let us see if we can run both motors. Obviously, we have to send DIR (direction) and PWM (power) commands for both motors A and B, so the following statements in the `loop()` function would run both motors:

```
digitalWrite(DIRB, CW);
analogWrite(PWMB, 150);
//and similarly for A:
digitalWrite(DIRA, CW);
analogWrite(PWMA, 150);
```

This code should make both motors run in the same direction at the same speed. In fact, this means that the entire robot should move forward (or backward, depending on how you wired the motors). Then, to stop both motors, you would write:

```
analogWrite(PWMA, 0);
analogWrite(PWMB, 0);
```

But now look at your code. It is quite bewildering for a newcomer who reads it. In fact, if you read the code six months later, it will make no sense. The method for making your code readable is to use functions and function calls. Any time you need to repeat a body of code, you can make a function with a descriptive name, and then to repeat that body of code, you call the function with appropriate values of the parameters.

The forward() Function

For example, if you want to make the robot move forward with power 150, you could do so with a simple command such as:

```
forward(150);
```

Remember that all our code lines end with a semicolon (;). But then, the forward statement is not understood by the Arduino. So you define a function called forward(), which contains the commands you already wrote to make the robot go forward. You should be able to run the same command with different speeds, such as forward(100), or forward (230). So you should define this function with a variable that represents the speed.

Here is the function definition:

```
void forward (byte spd)
{
      digitalWrite(DIRB, CW);
      analogWrite(PWMB, spd);
//and similarly for A:
      digitalWrite(DIRA, CW);
      analogWrite(PWMA, spd);
}
```

Since we want to move forward, the directions for both motors are fixed CW. The power for each motor is specified by the variable spd, which is defined in the very first line of code, within the parentheses, where it is stated that spd is a variable of type byte, which means that it can take only values from 0 to 255.

The function definition is given separately, outside the loop() function. When you call the function from within the loop, you call it with a specific value for the variable, as in forward(100).

The word "void" before the function simply states that the function does not return, or give, a value. It simply *does* something; in this case, it turns the motors at a specified speed.

In fact, you have already seen the setupArdumoto() function in action. It initializes the pins and is called at the very beginning of the program. If you read the code in order, the function call setupArdumoto() at the beginning makes no sense until you go down and see the function definition (what it does) after the loop() function. No parameter values need to be given to this function, so the parentheses are empty in the function definition and in the function call.

The delay() Function

The `delay()` function is a built-in function in the Arduino library. If you call this function, for example with a parameter value of 1000, the program does not proceed to the next statement for a time of 1000 milliseconds, or one second.

The Turns

How do you make the robot turn left or right? The way to accomplish this is by holding one wheel and turning the other. So the following lines of code would make a left or right turn, depending on which of the two motors, A or B, is on the right.

```
Void turnRight(byte spd)
{
     digitalWrite(DIRA, CW);
     analogWrite(PWMA, spd);
     analogWrite(PWMB, 0);
}
```

In the `loop()`, `turnRight(120)` would make the robot turn right by running MOTOR_A at the speed of 120, while stopping MOTOR_B.

Code for the Ultrasonic Sensor

Open the IDE, and type the following block of code.

Before you can use the code, you must add the library for the ultrasonic sensor to the Arduino environment on your computer. This is the library NewPing.h, which is available from the web. Type "NewPing.h" into Google and download the library (a zip file) into your computer following the instructions at the Arduino website: http://arduino.cc/en/Guide/Libraries#.UwjwKfldW7o

Open the IDE and type the following code into the sketch.

```
#include <NewPing.h>

#define TRIGGER_PIN  7   // Arduino pin tied to trigger pin on the ultrasonic sensor.
#define ECHO_PIN     6   // Arduino pin tied to echo pin on the ultrasonic sensor.
#define MAX_DISTANCE 200 // Maximum distance we want to ping for (in centimeters). Maximum
                         // sensor distance is rated at 400-1000cm.

NewPing sonar(TRIGGER_PIN, ECHO_PIN, MAX_DISTANCE); // NewPing setup of pins and maximum
                                                    // distance.

void setup() {
  Serial.begin(9600); // Open serial monitor at 9600 baud to see ping results.
}
```

```
void loop() {
  delay(50);
  unsigned int uS = sonar.ping(); // Send ping, get ping time in microseconds (uS).
  Serial.print("Ping: ");
  Serial.print(uS / US_ROUNDTRIP_CM); // Convert ping time to distance in cm and print
                                      // result (0 = outside set distance range)
  Serial.println("cm");
if ( uS / US_ROUNDTRIP_CM > 10 || uS / US_ROUNDTRIP_CM == 0) {
//   forward();
  Serial.println("Forward Ho!"); // This is only for fun!
}
else if (uS / US_ROUNDTRIP_CM < 10) {
  //stop();
  Serial.println("Stop, Stop!!"); //This line is also just for fun!
  delay(500);
  delay(500); */
}
}
```

How the Code Works

The first three lines define variables named TRIGGER_PIN, ECHO_PIN, and MAX_DISTANCE that have specific functions as described below.

The variable TRIGGER_PIN refers to the pin on the Arduino/Ardumoto board that is connected to the Trig pin on the ultrasonic sensor. In this case, the value is set to 7, which means that the digital pin number 7 on the Arduino board is connected to the Trig pin on the sensor, using the breadboard and jumpers. We could have connected the Trig pin to pin number 4 on the Arduino board, in which case we would have set this variable to 4.

The variable ECHO_PIN refers to the pin on the Arduino board that is connected to the Echo pin on the ultrasonic sensor. In this case, we set the value to 6, which is the pin number on the Arduino board connected to the Echo pin on the sensor, using the breadboard and jumpers. We could have connected the Echo pin to pin number 5 on the Arduino board, in which case we would have set this variable to 5.

MAX_DISTANCE is the variable that specifies the maximum distance that the ultrasonic sensor will be capable of detecting objects. In this case, if the distance from the sensor to the object is ≤200 cm, the sensor can measure and find the object; if the object is beyond 200 cm, the sensor will not be able to detect the object, and will show 0.

The next statement creates a new object named `sonar` and the type of that object is `NewPing`. This object takes three parameters, which are precisely the three variables defined earlier: `TRIGGER_PIN`, `ECHO_PIN`, and `MAX_DISTANCE`.

The next two declarations are two functions: `setup` and `loop`. The `setup` function opens the serial monitor. The serial monitor enables communication between the Arduino and your computer, while they are connected via the USB cable. The loop is the meat of the program. This function tells the Arduino what to do, in an endless loop, which continues until you shut down the robot.

The next line, `unsigned int uS = sonar.ping()`, measures the time between the signal pulse from the sensor to the object and the return pulse from the object in microseconds.

The next two lines convert the microseconds to distance in centimeters by using the functions available in the NewPing library.

The processor also performs another test. It checks whether the distance from the robot to the object is greater than 10 cm or equal to 0 (which means that the distance is greater than the maximum distance of 200 cm stipulated earlier), and if that is the case, it prints a message "Forward." Otherwise, if the distance is less than 10 cm, it prints "Stop."

PART 3: PUTTING IT ALL TOGETHER

Now you can understand the entire code in a program that makes the robot roam around a room, avoiding obstacles. In order to write such a program, you should first plan each step. Describe what you want the robot to do. For example, we might write down something like this:

1. Move forward at some speed, all the time pinging to measure the distance to the wall in front of the robot.

2. If the distance is greater than some value (to be decided by you) continue to move forward.

3. If the distance is less than the value, turn right (or left).

4. Go back to instruction 1 and loop.

Note that each of the tasks above has already been tried out by you. So the total code should be easy to understand.

To make the robot productive and have it sweep the floor, attach a piece of a dust cloth (such as a Swiffer cloth) to the robot in front, as shown in Figure 2.8.

Figure 2.8
Dust cloth attached to the robot.

Now the robot moves around the room in random fashion, avoiding the walls and other obstructions, all the while sweeping the floor, and you have a simple version of something like the Roomba!

Complete Code for the Sweeper Robot

The software written for this project must ensure that the robot avoids obstacle as it roams around an area and sweeps up the dust in its path. The following instructions achieve the purpose of the project.

```
// Arduino-Ardumoto obstacle avoiding robot

#include <NewPing.h>

#define TRIGGER_PIN  7    // Arduino pin tied to trigger pin on the ultrasonic sensor.
#define ECHO_PIN     6    // Arduino pin tied to echo pin on the ultrasonic sensor.
#define MAX_DISTANCE 200  // Maximum distance we want to ping for (in centimeters). Maximum
                          // sensor distance is rated at 400-1000cm.

#define CW   0
#define CCW 1
```

```
// Motor definitions to make life easier:
#define MOTOR_A 0
#define MOTOR_B 1

// Pin Assignments
// Don't change these! These pins are statically defined by the shield layout
const byte PWMA = 3;   // PWM control (speed) for motor A
const byte PWMB = 11;  // PWM control (speed) for motor B
const byte DIRA = 12;  // Direction control for motor A
const byte DIRB = 13;  // Direction control for motor B

NewPing sonar(TRIGGER_PIN, ECHO_PIN, MAX_DISTANCE); // NewPing setup of pins and maximum
                                                    // distance.

void setup()
{
  setupArdumoto(); // Set all pins as outputs
}

void loop()
{
  delay(50);
  unsigned int uS = sonar.ping(); // Send ping, get ping time in microseconds (uS).
  if ( uS / US_ROUNDTRIP_CM > 50 || uS / US_ROUNDTRIP_CM == 0)
  {
  //   Move forward
  forward(150);
  }
  else if (uS / US_ROUNDTRIP_CM < 50)
  {
  turnRight(100);
  delay(500);
  }
}
// driveArdumoto drives 'motor' in direction 'dir' at speed 'spd'
void driveArdumoto(byte motor, byte dir, byte spd)
{
  if (motor == MOTOR_A)
  {
    digitalWrite(DIRA, dir);
    analogWrite(PWMA, spd);
  }
  else if (motor == MOTOR_B)
```

```
  {
    digitalWrite(DIRB, dir);
    analogWrite(PWMB, spd);
  }
}

void forward(byte spd)      // Runs both motors at speed 'spd'
{
  driveArdumoto(MOTOR_A, CW, spd);      // Motor A at speed spd
  driveArdumoto(MOTOR_B, CW, spd);      // Motor B at speed spd
}

void turnRight(byte spd)
{
  stopArdumoto(MOTOR_B);                //Motor B stop
  driveArdumoto(MOTOR_A, CW, spd);      //Motor A run
}

// stopArdumoto makes a motor stop
void stopArdumoto(byte motor)
{
  driveArdumoto(motor, 0, 0);
}



// setupArdumoto initializes all pins
void setupArdumoto()
{
  // All pins should be set up as outputs:
  pinMode(PWMA, OUTPUT);
  pinMode(PWMB, OUTPUT);
  pinMode(DIRA, OUTPUT);
  pinMode(DIRB, OUTPUT);

  // Initialize all pins as low:
  digitalWrite(PWMA, LOW);
  digitalWrite(PWMB, LOW);
  digitalWrite(DIRA, LOW);
  digitalWrite(DIRB, LOW);
}
```

How the Code Works

As you can see, this code is an extension of the previous sketch in this chapter. Additional code and explanations are given below.

The beginning of the code defines a set of variables for the trigger pin, echo pin, and max distance (to measure the distance from the trigger to the object), CW 0 and CW 1 to turn the motors clockwise and counter-clockwise, and motor A and motor B for the two motors used by the robot.

The next set of definitions is for constants that are connected to the various pins on the shield. PWMA is connected to pin 3, PWMB is connected to pin 11, DIRA is connected to the pin 12, and DIRB is connected to the pin 13. You should have already done these connections while you were constructing the robot.

The `NewPing` statement creates the sonar object.

In the setup procedures, the Ardumotor board is set up with all pins configured as outputs. Check the `setupArdumoto()` function at the bottom of this code.

In the `void loop()` function, the following activities are accomplished: `delay(50)` holds, or halts, for 50 milliseconds before starting the loop every time (repeating the sequence of actions in the loop).

For the next two lines of the code (`unsigned int ...`), refer to the previous block of code for using the ultrasonic sensor.

The best way to understand the code for the motor control, which is the next section of the code block, is to start after the line:

```
void loop()
{
........
}
```

and look at all the functions defined there. The `loop()` function tells the processor what to do while running the program. These instructions are given in the form of function calls. For example, look at the line:

```
forward (150);
```

in the code of the `loop()` function. This statement calls the function `forward()` with a certain value within the parentheses, in this case, 150. The parameter is a variable named `spd` of type byte, which can have values in the range (0 ... 255). The function forward(byte spd) is defined after the loop() and driveArdumoto() functions. The code is repeated here:

```
void forward(byte spd)      // Runs both motors at speed 'spd'
{
  driveArdumoto(MOTOR_A, CW, spd);      // Motor A at speed spd
  driveArdumoto(MOTOR_B, CW, spd);      // Motor B at speed spd
}
```

This function tells the processor what should be done whenever the function forward() is called with a particular value for the parameter spd. It turns both motors clockwise with the same speed.

The next statement uses the if condition to check if the robot is greater than 50 cm from the wall, and if so it will move forward; otherwise if the robot is closer than 50 cm, then the robot turns right for 500 milliseconds, which is the given delay. Check for the turnRight() function in the functions at the bottom.

The next driveArdumoto() function drives this specified motor in the specified direction at the specified speed. This function takes three parameters, which specify the motor, direction, and speed.

If motor = MOTOR_A, then the digitalWrite command (which has two parameters: a pin number on the Arduino board and dir, which is a byte value) places the value of the variable dir on that pin number. In this case, it is pin number 12 as was assigned in the pin assignments part of the code.

The analogWrite command places the value of the second parameter speed on the assigned pin PWMA, which is pin number 3 as assigned earlier.

If motor = MOTOR_B, then the same command places value numbers 13 for DIRB and 11 for PWMB.

The function forward was already explained in the motor code in the earlier example.

The function void turnRight takes one parameter called speed of type byte. In the function, stopArdumoto(MOTOR_B) stops motor B. The statement driveArdumoto takes three parameters, which are the motor, the direction, and the speed. Using these values, the robot turns appropriately.

The function setupArdumoto sets up specific pins of the Arduino board that are used in this program as input pins or output pins. It also initializes these pins to be either low or high. In this case, the pin numbers 3, 11, 12, and 13, were assigned earlier to speed control of motor A (PWMA), speed control for motor B (PWMB), direction control of motor A (DIRA), and direction control of motor B (DIRB). These pins are also initialized to LOW or with a 0 value.

You can now test your robot by preparing a small enclosure with cardboard boxes and leaving your robot in that area. Notice that the dust cloth in fact works: look at the dust on the dust cloth as the robot moves on the floor.

CONCLUSION

In this chapter you have learned how to design a robot that can avoid obstacles. You have also experimented with the concepts and designed a project that has real-life applicability and possibly a commercial relevance. You can further expand on this project by using more sophisticated artificial intelligence algorithms to design a more sensitive and functional robot vacuum cleaner.

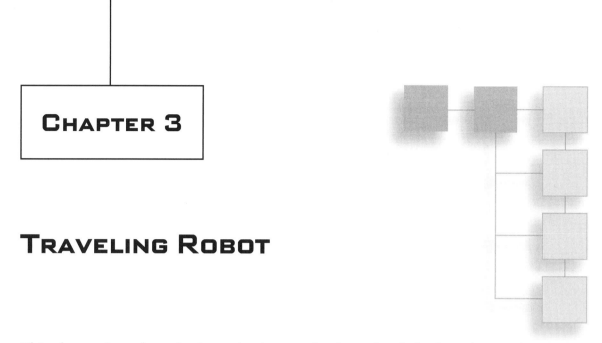

CHAPTER 3

TRAVELING ROBOT

This chapter introduces basic navigation mechanisms that help the robot navigate using line tracking. You will also learn how the robot can detect colors along its path. This feature can become a good option for making the robot take different paths depending on the color detected.

CHAPTER OBJECTIVES

- Learn how an IR line sensor works
- Use IR line sensors and make a robot follow a track
- Learn how a color sensor works
- Use a color sensor and make a robot perform specific tasks while following a track

INTRODUCTION

This chapter is organized in four parts:

Part 1: Mount the line sensors on the robot chassis, learn how the line sensors work, and get appropriate output numbers for your flooring and track.

Part 2: Use the data obtained in Part 1 to design a lineTracker sketch that makes the robot follow the track.

Part 3: Mount the color sensor on the robot chassis, connect it to the Arduino board, learn how the sensor works, and get output numbers for different colors.

Part 4: Use the data obtained in Part 3 to add functionality to the robot by programming it to carry out different tasks when it detects a different color.

MATERIALS REQUIRED

- QRE1113 Analog line sensors (quantity: 2)
- Robot chassis with Arduino board and Ardumoto motor control shield (as built and used in Chapter 2)
- TCS34725 color sensor

PART 1: LINE SENSOR

For this project, you need to use a line sensor to guide the robot to follow a line.

How Does a Line Sensor Work?

Using a line sensor is simple enough. On the robot chassis (which you built in Chapter 2), you mount two line sensors. Each line sensor has an IR emitter-detector pair. The sensors are mounted so that the pair is close to the floor. Lay a track on the floor using black electrical tape as shown in Figure 3.1.

Figure 3.1
Line sensor track.

The black color is suitable if the floor is light colored. If the floor itself is dark, you can lay the track with white tape. There must be good contrast between the track and the floor. The line sensors are mounted close together in front of the chassis (Figure 3.2). If the chassis is centered on the track, both line sensors give out the same (or nearly the same) signal, and the robot moves forward. If the robot deviates a little from the track, one sensor gives a different signal from the other, and the robot corrects itself.

Figure 3.2
Mounted line sensors.

The line sensor we used is a QRE1113 Analog line sensor, available from SparkFun. It has three pins: GND, VCC, and OUT. The GND pin is connected to the Arduino GND pin, and the VCC is connected to 5V. The OUT pin gives an analog signal between 0 and 1023, depending on the amount of IR light reflected back to the detector. Light-colored surfaces reflect more IR, and the signal is a low number, whereas dark surfaces reflect less light, and therefore the signal is a high number.

Assembling the Robot with the Sensors

Start with the robot chassis built in the previous chapter. You will need to mount two line sensors on the chassis. First, solder three header pins to each sensor. Make sure that the longer legs of the header pins are facing away from the side on which the detector is mounted. After soldering, the sensor looks like Figure 3.3. Three jumper wires connect the header pins to the breadboard on top of the chassis. The sensor itself can be positioned on the chassis with bent paper clips, so that the detector is close to (about a quarter of an inch) and facing the floor.

Figure 3.3
Sensors with pins.

Now refer back to Figure 3.2 to make sure that the sensors are properly mounted in front of the chassis. The distance between the floor and the detector should be only about a quarter inch. The sensors must be close together. If necessary, adjust the paper clips to make sure you get them positioned closely.

Figure 3.4 shows the jumper wire connections.

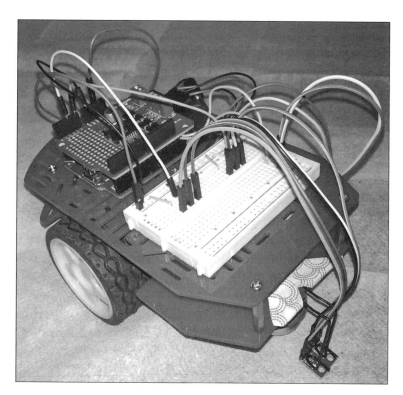

Figure 3.4
Jumper wire connections.

Sensor Pin	Arduino Pin
GND (both sensors)	GND
VCC (both sensors)	5V
OUT (left sensor)	A0
OUT (right sensor)	A2

lineSensorChk Sketch

Type the sketch given below and save it as "lineSensorChk" in the Arduino IDE.

```
/* lineSensorChk
   Two QRE1113 Analog line sensors from SparkFun connected to Analog pins 0 and 2
   Outputs to the serial monitor - Lower numbers mean light surface, more light reflected
*/
```

```
int qreLeft = A0;        //Left sensor OUT pin connected to Analog Pin 0 of Arduino
int qreRight = A2;       //Right sensor OUT pin connected to Analog Pin 2 of Arduino

void setup()
{
  Serial.begin(9600);         // Start the serial monitor
}
void loop()
{
  int QRE_Left = analogRead(qreLeft);      //Read the output value from the left sensor
  int QRE_Right = analogRead(qreRight);    // Read the output value from the right sensor
  Serial.print("left: ");                  // Start printing the values to the serial monitor
  Serial.print(QRE_Left);
  Serial.print("  right: ");
  Serial.println(QRE_Right);
  delay(2000);
}
```

Using the lineSensorChk Sketch

You will use the sketch to measure the outputs from each sensor when it is on top of the black tape and when it is on top of the lighter colored floor. Follow the steps given below.

Place the robot so that both sensors are right on top of the black tape. Start running the sketch. You will see the serial monitor display the output values from the right and left sensors. The monitor will show lines like this:

left: 960 right: 970

left: 960 right: 970

……………………..

These numbers are for illustrative purposes. You will get different numbers, but both pairs of numbers should be approximately the same. Write down one pair of the values on a sheet of paper.

While the sketch is running, turn the robot (with your hand) slightly to the left, so that the left sensor is facing the flooring, and the right sensor is still on the black tape. Now the monitor should show data similar to:

left: 150 right: 960

left: 160 right: 970

…………………..

Again, these numbers are examples. Write down one pair of the values.

Let the sketch continue to run, and now turn the robot the other way with your hand, so that the right sensor is facing the flooring, and the left sensor is on the black tape. The monitor should show something like

left: 960 right: 165

left: 960 right: 165

…………………..

It is not necessary to write down these values. Notice that there is a large gap between the values from the dark tape and those from the light flooring. If the contrast between track and flooring is not high, these two values will be closer. As long as there is a clear demarcation, and there is no overlap among the values, the project will work fine.

Choose a boundary value that clearly demarcates the high values from the dark tape and the low values from the light flooring. With data such as in the previous examples, you could easily choose 500 or 400 as the demarcation. You will use this value when you design the lineTracker sketch in Part 2 of this chapter.

How the lineSensorChk Code Works

The two integer variables refer to Analog pins 0 and 2.

The setup() function starts the serial communication with a baud rate of 9600.

The loop() function does the following:

Creates an integer variable QRE_Left that takes the output from the left IR sensor.

Creates an integer variable QRE_Right that takes the output from the right IR sensor.

The next four lines print the values to the serial monitor.

Loops back every two seconds.

PART 2: LINE TRACKING

Once the output values of the line sensors are obtained from the dark track and the light floor, you are ready to write the sketch that will make the robot follow the track. You can lay the track around the floor in a large room or on a table. You can use black insulation tape on a light-colored (e.g., maple) wooden floor. You could also use the black tape on a sheet of white school project board. Both options work fine.

LineTracker Sketch

Open the Arduino IDE and type the following code.

```
/* Line Tracker
    QRE1113 line sensors from SparkFun connected to Analog pins 0 and 2
    The line sensors are individual sensors
   The motor control commands are the same as those in the previous chapter
   The line sensor commands are from the previous activity
*/
#define CW 0
#define CCW 1

// Motor definitions:
#define MOTOR_A 0
#define MOTOR_B 1

// Pin Assignments //
// Don't change these! These pins are statically defined by Arduino & Ardumoto shield layout
const byte PWMA = 3;   // PWM control (speed) for motor A
const byte PWMB = 11;  // PWM control (speed) for motor B
const byte DIRA = 12;  // Direction control for motor A
const byte DIRB = 13;  // Direction control for motor B

int qreLeft = 0; //connected to Analog 2
int qreRight = 2; //connected to Analog 0

byte spd = 100; // forward speed
byte hiSpd = 120;   //turning speed

int threshold = 700;
// threshold for line sensor values. Greater than this means on the line (dark)
// less than this means the sensor is off the line (light)
// The threshold must be determined experimentally for each surface and track as done in the
// lineSensorChk sketch

void setup()
{
  setupArdumoto(); // Setup & initialize all motor drive pins
}

void loop()
{
  int QRE_Left = analogRead(qreLeft);    // Analog output from left sensor
  int QRE_Right = analogRead(qreRight);  // Analog output from right sensor
```

```
  if (QRE_Left > threshold && QRE_Right > threshold)
 //If both sensors are on the track, move forward
  {
    forward();
  }
  else if (QRE_Left < threshold && QRE_Right > threshold)    // if left sensor is off-track,
                                                              // turn right

  {
    bearRight();
  }
  else if (QRE_Left > threshold && QRE_Right < threshold)    // If right sensor is off-
                                                             // track, turn left

  {
    bearLeft();
  }
}
// driveArdumoto drives 'motor' in direction 'dir' at speed 'spd'
void driveArdumoto(byte motor, byte dir, byte spd)
{
  if (motor == MOTOR_A)
  {
    digitalWrite(DIRA, dir);
    analogWrite(PWMA, spd);
  }
  else if (motor == MOTOR_B)
  {
    digitalWrite(DIRB, dir);
    analogWrite(PWMB, spd);
  }
}

void forward()      // Runs both motors at speed 'spd'
{
  driveArdumoto(MOTOR_A, CW, spd);    // Motor A at speed spd
  driveArdumoto(MOTOR_B, CW, spd);    // Motor B at speed spd
}

void bearRight()
{
  driveArdumoto(MOTOR_B, CW, 0);        //Motor B STOP
  driveArdumoto(MOTOR_A, CW, hiSpd);    //Motor A hiSpd
}
```

```
void bearLeft()
{
  driveArdumoto(MOTOR_B, CW, hiSpd);      //Motor B hiSpd
  driveArdumoto(MOTOR_A, CW, 0);      //Motor A STOP
}



// setupArdumoto initializes all pins
void setupArdumoto()
{
  // All pins should be set up as outputs:
  pinMode(PWMA, OUTPUT);
  pinMode(PWMB, OUTPUT);
  pinMode(DIRA, OUTPUT);
  pinMode(DIRB, OUTPUT);

  // Initialize all pins as low:
  digitalWrite(PWMA, LOW);
  digitalWrite(PWMB, LOW);
  digitalWrite(DIRA, LOW);
  digitalWrite(DIRB, LOW);
}
```

How the Code Works

The statements for the motor controls (define CW 0, CW 1, Motor_A 0 and Motor_B 1) have been discussed in Chapter 2, "Build Your Own Robot Sweeper." These are statements that control the motors.

The pin assignments section has also been discussed in Chapter 2. These pins control the speed and direction of the motors.

The qreLeft and qreRight are the line sensors and are connected to Analog pins A0 and A2 on the Arduino board.

The next two statements, spd and hiSpd, are used to control the forward speed and turning speed of the robot.

Output values of the sensors have been explained in the lineTracker sketch, where values higher than 700 in this case refer to a darker color on the line.

The setup() function initializes all the motor drive pins.

The loop() function does the following:

Reads the values from the left and right line sensors and assigns them to the variables QRE_Left and QRE_Right.

If QRE_Left is > threshold value and QRE_Right is > threshold value, the robot moves forward.

Otherwise if QRE_Left is < threshold and QRE_Right is > threshold, the robot bears right.

Otherwise if QRE_Left is > threshold and QRE_Right is < threshold, the robot bears left.

The driveArdumoto(byte motor, byte dir, byte spd) function was also discussed in Chapter 2. It runs a specified motor in a specified direction at a specified speed.

The forward() function runs both the motors at the same value of spd in a clockwise motion so the robot moves forward.

The bearLeft() function drives MOTOR_A with the value of hiSpd, and stops MOTOR_B.

The bearRight() function drives MOTOR_B with the value of hiSpd, and stops MOTOR_A.

The setupArdumoto() function initializes all the pins as OUTPUT or LOW.

You can get a lot of entertainment with the lineTracker sketch. You can make the robot run all around the house along a specific track. We tried a short 8-foot track. When the robot went past the end, it kept turning until it found the track again and followed it back to the starting point.

PART 3: LEARNING TO USE THE COLOR SENSOR

Now, can we make this line tracker do something more? For example, can we make it stop somewhere, or follow a separate track or something like that? Well, first we should make sure the robot recognizes where we want it to stop. So, we put a small strip of a different color tape at one point, by the side of the track. We tried a red strip (Figure 3.5).

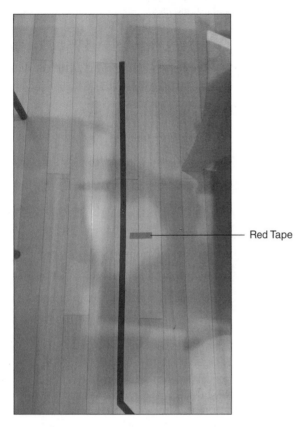

Red Tape

Figure 3.5
Additional color tape on the track.

To recognize the strip, we mounted a color sensor (TCS34725) on one side of the robot, again facing the floor and close to it (see Figure 3.6).

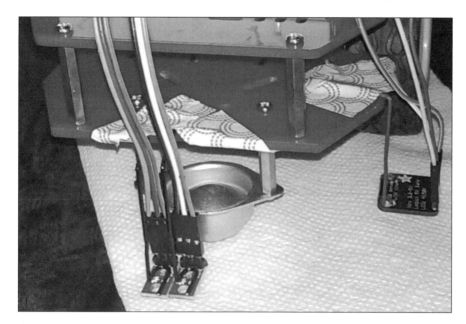

Figure 3.6
Color sensor mounted on the robot.

The color sensor, shown in Figure 3.7, has 8 pins to which you should first solder header pins, as you did for the IR line-tracker sensors in Part 1.

Figure 3.7
Header pins on the color sensor.

This color sensor does not have an output pin that can be connected to either the Analog or the Digital pins on the Arduino. Instead, it has a Serial Clock (SCL) and a Serial Data (SDA) pin, which are used by the Arduino I2C serial bus to read the data from the sensor (Figure 3.8).

Figure 3.8
Pins on the color sensor.

I2C is a standardized serial connection between integrated circuits, involving only two wires (SCL and SDA). On the Arduino, these two connections are made by using two pins, which are on the same side of the board as the Digital pins. After Digital pins 0–13, you will see, in order, GND, AREF, SDA, and SCL (Figure 3.9).

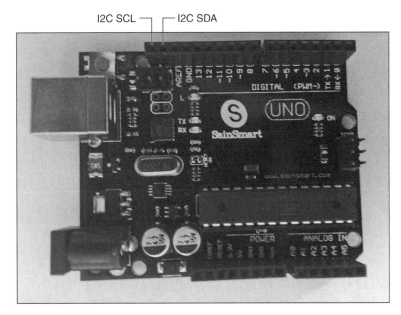

Figure 3.9
Pins on the Arduino board.

Connections between the Color Sensor Pins and Arduino Board Pins

Use the following table and make the appropriate connections between the pins on the color sensor and pins on the Aduino board.

TCS34725 Pin	Arduino Pin
GND	GND
Vin	5V
SDA	SDA
SCL	SCL

The Ardumoto board, which is driving the motors, comes in the way of making the I2C connections. Unfortunately, the Ardumoto board does not provide header pins for the I2C connections. So you will have to take the SCL and SDA wires from under the Ardumoto board as shown in Figure 3.10

Figure 3.10
Accessing the SCL and SDA wires.

The sensor outputs values for R (Red), G (Green), B (Blue), and C (Clear). You should download two libraries: Wire.h and Adafruit_TCS34725.h. The first library is for the I2C bus control, and the second one is to control the color sensor.

lineClrSensor Sketch for Testing the Line Sensors and the Color Sensor

Type the sketch lineClrSensorChk in the Arduino IDE. You will use the sketch to measure the outputs from the two IR sensors when each is on top of the black tape, and on top of the lighter-colored floor, and when the color sensor is on top of the black tape, on the flooring, and on the red colored strip.

Open the Arduino IDE and type the following sketch.

```
/*  line sensor & color sensor tester
    QRE1113 line sensors from SparkFun connected to Analog pins 0 and 2
    The line sensors are individual sensors, used to track the main line
    TCS34725 color sensor used to STOP at a point on the track
    Color sensor connected to the I2C bus on the Arduino
    */
```

```
#include <Wire.h>      //Library for communicating with I2C device
#include <Adafruit_TCS34725.h>     //Library of commands for the color sensor

/* Initialize instance of color sensor with default values (int time = 2.4ms, gain = 1x) */
Adafruit_TCS34725 tcs = Adafruit_TCS34725();

int qreLeft = A0; //connected to Analog 0
int qreRight = A2; //connected to Analog 2

void setup(){
  Serial.begin(9600);
  if (tcs.begin())
  {
  }
  else
  {
    Serial.println("Give it up!");
    while (1);
  }
}


void loop(){
    uint16_t r, g, b, c, colorTemp, lux;      //Declare output variables for color sensor
    int QRE_Left = analogRead(qreLeft);    //Output of left IR sensor
    int QRE_Right = analogRead(qreRight);     // Output of right IR sensor
    tcs.getRawData(&r, &g, &b, &c);            // Read the color sensor outputs

    //Print values to the serial monitor
    Serial.print("R: "); Serial.print(r); Serial.print("  G: "); Serial.print(g);
    Serial.print("  B: "); Serial.print(b); Serial.print("  C: "); Serial.println(c);
    Serial.print("  left: "); Serial.print(QRE_Left);
    Serial.print("  right: "); Serial.println(QRE_Right);
    Serial.println("");
    delay(2000);

}
```

Testing the lineClrSensor Sketch

Follow these steps to test your sketch.

1. Place the robot so that both line sensors are right on top of the black tape, and the color sensor is facing the light-colored floor. Start running the sketch. You will see

the serial monitor display the output values from the right and left sensors, and from the color sensor. The monitor will show lines like this:

R: 10, G: 15, B: 12, C: 40

left: 960 right: 970

………………..

These numbers are for illustrative purposes. You will get different numbers. The color sensor values for RGBC will be much lower than those for the two line sensors. Write down a set of the values on a sheet of paper.

2. While the sketch is running, turn the robot (with your hand) slightly to the left, so that the left sensor is facing the flooring, and the right sensor is still on the black tape. Now the monitor should show data such as:

R: 10, G: 15, B: 12, C: 50

left: 150 right: 960

Again, these numbers are examples. The RGBC values won't change much. Write down this set of values.

3. Let the sketch continue to run, and now turn the robot the other way with your hand, so that the right sensor is facing the flooring, and the left sensor is on the black tape. The monitor should show something like:

R: 50, G: 15, B: 12, C: 50

left: 960 right: 165

………………..

It is not necessary to write down these values. Notice that there is a large gap (as in the sketch for Part 1) between the values of the line sensors from the dark tape and those from the light flooring.

4. With the sketch continuing to run, place the robot so that the line sensors are on the black tape and the color sensor is on the red tape marker. Now you should see a large difference in the color sensor values—either the R value or the C value. You can use whichever value shows a good change, so you can choose a boundary value that demarcates the high value from the red marker and the low value from the floor and black tape. Choose such a boundary value. With data such as provided above, you could easily use the red (R) value and choose 30 or 40 as the demarcation. You will use this value when you design the lineClrSensor sketch in Part 4 of this chapter.

How the Code Works

The include files ensure the proper references to the required libraries have been established in the sketch. Wire.h controls the I2C bus, and the Adafruit_TCS4725.h controls the color sensor.

Next, the color sensor is initialized with default values.

The variables qreLeft and qreRight are connected to Analog pins A0 and A2.

The setup() function checks to see if the color sensor is ready. If not, an error message is printed.

The loop() function does the following:

A set of variables are used to hold the output values from the color sensor.

The values from the Analog pins are stored in QRE_Left and QRE_Right. .

The raw data from the color sensor are queried.

The next set of statements prints the values stored in the variables for the RGBC fr the color sensor and from QRE_Left and QRE_Right.

A delay of 2000 ms is used to pause the loop for two seconds and start executing the statements in the loop again.

PART 4: MAKING THE ROBOT FOLLOW A TRACK AND STOP AT A SPECIFIC POINT FOR A PRESCRIBED TIME

Once you know the values of the outputs from the line sensors for the black tape and the floor, and the output of the color sensor for the floor and the red stripe, you are ready to design the sketch. We used the R value, but you can use any of the RGBC values that have a clear demarcation. Also note that the commands for the line sensors and the motor controls do not change from Part 2.

lineandColorSensorTest Sketch

Open your IDE and type the following sketch and save it as lineandColorSensorTest

```
/* Line and color tracker tester
   by: Chandra Prayaga Feb 16, 2014
   QRE1113 line sensors from SparkFun connected to Analog pins 0 and 2
   The line sensors are individual sensors*/
```

```
//Code for the QRE113 Analog board
//Outputs via the serial terminal - Lower numbers mean more reflected light

#include <Wire.h                 // Control I2C bus
#include "Adafruit_TCS34725.h"     //Control color sensor

/* Initialize with default values (int time = 2.4ms, gain = 1x) */
Adafruit_TCS4725 tcs = Adafruit_TCS34725();

#define CW
#define CC

// Motor  initions:
#define  OR_A 0
#defin   TOR_B 1

// Pi  signments //
// D  change these! These pins are statically defined by Arduino & Ardumoto shield layout
con   yte PWMA = 3;   // PWM control (speed) for motor A
co    byte PWMB = 11;  // PWM control (speed) for motor B
c     byte DIRA = 12;  // Direction control for motor A
      byte DIRB = 13;  // Direction control for motor B

qreLeft = A0; //connected to Analog 2
  qreRight = A2; //connected to Analog 0

yte spd = 100; // forward speed
byte hiSpd = 120;   //turning speed

int threshold = 700;
// threshold for line sensor values. Greater than this means on the line (dark)
// less than this means the sensor is off the line (light)
// This must be determined experimentally for each surface and track

void setup()
{
  setupArdumoto(); // Set up & initialize all motor drive pins

  if (tcs.begin())
  {
  }
  else
  {
    allStop();
    while (1);
  }
}
```

```
void loop()
{
  uint16_t r, g, b, c, colorTemp, lux;
  int QRE_Left = analogRead(qreLeft);
  int QRE_Right = analogRead(qreRight);
  tcs.getRawData(&r, &g, &b, &c);

  if (r > 40)        //if color sensor is on the bare floor
  {
    if (QRE_Left > threshold && QRE_Right > threshold)
    {
      forward();
    }
    else if (QRE_Left < threshold && QRE_Right > threshold)
    {
        bearRight();
    }
    else if (QRE_Left > threshold && QRE_Right < threshold)
    {
        bearLeft();
    }
  }
 else       //If color sensor is on the red tape marker
  {
  allStop();
  delay(5000);        //Stop as long as you want - in this case, 5 seconds
  forward();
  delay(100);
  }
}
// driveArdumoto drives 'motor' in direction 'dir' at speed 'spd'
void driveArdumoto(byte motor, byte dir, byte spd)
{
  if (motor == MOTOR_A)
  {
    digitalWrite(DIRA, dir);
    analogWrite(PWMA, spd);
  }
  else if (motor == MOTOR_B)
  {
    digitalWrite(DIRB, dir);
```

```
      analogWrite(PWMB, spd);
   }
}

void forward()     // Runs both motors at speed 'spd'
{
  driveArdumoto(MOTOR_A, CW, spd);     // Motor A at speed spd
  driveArdumoto(MOTOR_B, CW, spd);     // Motor B at speed spd
}

void bearRight()
{
  driveArdumoto(MOTOR_B, CW, 0);       //Motor B Stop
  driveArdumoto(MOTOR_A, CW, hiSpd);   //Motor A hiSpd
}

void bearLeft()
{
  driveArdumoto(MOTOR_B, CW, hiSpd);   //Motor B hiSpd
  driveArdumoto(MOTOR_A, CW, 0);     //Motor A Stop
}

// stopArdumoto makes a motor stop
void stopArdumoto(byte motor)
{
  driveArdumoto(motor, 0, 0);
}

void allStop()                 //Stop both motors
{
  stopArdumoto(MOTOR_A);   // Stop motor A

  // All pins should be set up as outputs:
  pinMode(PWMA, OUTPUT);
  pinMode(PWMB, OUTPUT);
  pinMode(DIRA, OUTPUT);
  pinMode(DIRB, OUTPUT);

  // Initialize all pins as low:
  digitalWrite(PWMA, LOW);
  digitalWrite(PWMB, LOW);
  digitalWrite(DIRA, LOW);
  digitalWrite(DIRB, LOW);
}
```

How the Code Works

The include files are the same as in Part 3.

Next, the color sensor is initialized with default values.

The next set of statements defines the motors and their movement.

The set of constants assigns the pins for controlling the speed and direction of the Ardumoto.

qreLeft and qreRight are variables assigned to the pins Analog 0 and Analog 2 on the Arduino board.

The next two variables of type byte set the values for forward speed and turning speed

An integer variable threshold is used to set a value of 700 against which all other color values will be tested.

The setup() function does the following:

Calls the setupArdumoto() function.

Checks if the color sensor is awake.

Otherwise it stops the motors with allStop() and stops the program from moving further with the while(1) statement that ensures that the program does not proceed further since the condition in the while loop is always satisfied.

The loop() function does the following:

A set of variables are used to hold the output values from the color sensor.

The values from the Analog pins are stored in QRE_Left and QRE_Right.

The raw data from the color sensor are also queried for r, g, b and c.

If QRE_Left > threshold and QRE_Right > threshold (both line sensors on black tape), the forward() function is called.

Otherwise if QRE_Left < threshold and QRE_Right > threshold (left sensor off track), the bearRight() function is called.

Otherwise if QRE_Left > threshold and QRE_Right < threshold (right sensor off track), the bearLeft() function is called.

Otherwise, (color sensor on red tape marker) the allStop() function is called, and a delay of 5000 ms is set before the loop starts again.

The `driveArdumoto(byte motor, byte dir, byte spd)` function does the following:

A check is used to check if the current motor is MOTOR_A, and then the direction is written to the DIRA pin, and speed is written to the PWMA pin.

Otherwise if the motor is MOTOR_B, then the direction is written on the DIRB pin, and speed is written on the PWMB pin.

The `forward()` function runs both motors at the specified speed by the value set in the variable `spd`.

The `bearRight()` function stops MOTOR_B and drives MOTOR_A with the value set in the variable `hiSpd`.

The `bearLeft()` function stops MOTOR_A and drives MOTOR_B with the value set in the variable `hiSpd`.

The `stopArdumoto(byte motor)` function makes a motor stop by giving it a value of `spd` 0.

The `allStop()` function stops both motors.

The `setupArdumoto()` function does the following:

Sets up all pins as outputs.

Initializes all pins as low.

CONCLUSION

In this chapter, you looked at how a line-tracking algorithm works using infrared sensors. You also looked at how to add a color sensor and use the values of `r`, `g`, and `b` for use in other tasks such as stopping a robot or following a different path.

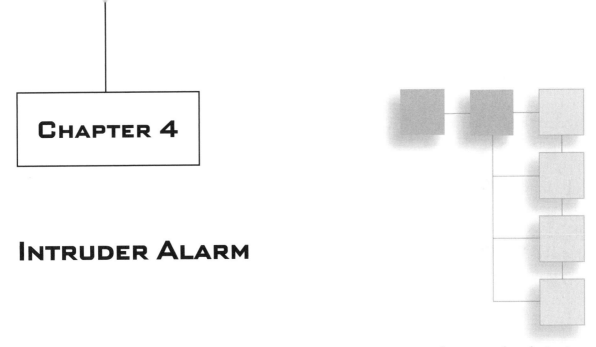

CHAPTER 4

INTRUDER ALARM

In this chapter, you will learn how to use various sensors as detectors for designing intruder alarms.

CHAPTER OBJECTIVES

- Use a laser diode and a photoresistor sensor to create an intruder alarm
- Use an ultrasonic range sensor to design a proximity alarm
- Use a capacitive touch sensor to detect an object being touched
- Use the capacitive touch sensor to create an LED display

INTRODUCTION

The basic principle behind each of the chapter objectives is described here:

Intruder alarm: A photoresistor is a simple sensor for detecting a laser beam. If you set up a diode laser and a photoresistor so that the laser beam falls on the photoresistor, the value of the resistance changes when the laser beam is cut—hence the name, photoresistor. This change in resistance can be detected by a circuit and used to sound an audio alarm or blink an alarm light.

Proximity alarm: You have already used the ultrasonic range sensor in Chapter 2 when you designed the obstacle-avoiding robot. The ultrasonic sensor is used to find the distance to an object within a specified range. You can set the range to a convenient distance so that if an object (or person) approaches within that distance, the system triggers an alarm (sound or light).

Touch sensor and alarm: If a person touches an object, the electrical capacitance of the human body changes the electrical characteristics of a circuit close to the object. These changes can be detected and an alarm produced.

Keyboard and LEDs: The same principle of detecting the capacitance of the human body can also be used to create a touch keyboard, which can then control LEDs or generate music.

MATERIALS REQUIRED

- Arduino board
- Laser diode sensor
- Photoresistor sensor
- Ultrasonic range sensor
- LEDs of different colors
- Solderless breadboard and hookup wires

ACTIVITY 1: INTRUDER ALARM WITH A DIODE LASER AND PHOTORESISTOR

Figure 4.1 shows the setup, including the diode laser and the photoresistor. The laser and the sensor are close together here, but you can imagine setting up the laser beam across a window, for example, so that if a burglar climbs through the window, he interrupts the laser beam and an alarm is generated.

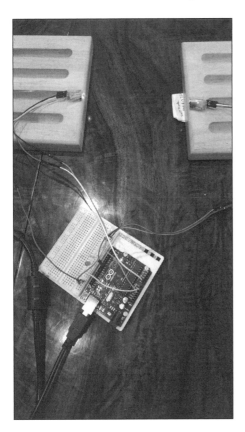

Figure 4.1
Intruder alarm setup.

The diode laser module we used is available as one of several Arduino-compatible sensors from Sunfounder in kit form, available through Amazon. The module has three pins, with the outer pins labeled "–" and "S" as shown in Figure 4.2.

Figure 4.2
Diode laser module.

The pin connections are shown in Table 4.1.

Table 4.1   Diode Laser Pin Connections

Diode Laser Pin	Arduino Pin
–	GND
Middle	5V
S	8 (Any digital pin will do the job)

Sketch to Control the Laser

The sketch presented below is an exercise to control the laser.

```
/*
Laser diode check
*/
void setup ()
{
  pinMode (8, OUTPUT); // Digital output to laser S pin
}
void loop ()
{
  digitalWrite (8, HIGH); // Laser ON
  delay (1000); // delay one second
  digitalWrite (8, LOW); // Laser OFF
  delay (1000); // delay one second
}
```

How the Code Works

This sketch demonstrates the use of the laser diode.

The setup() function sets the digital output from the Arduino pin 8 to the laser S pin.

The loop() function does the following:

Turns on the laser by setting it to HIGH.

Sets a delay of one second.

Turns the laser off by setting it to LOW.

Sets a delay of one second.

So for every one second, the laser turns on, remains on for a second, and then turns off and remains off for a second. This cycle is repeated in this function.

Sketch to Control the Photoresistor

Let's now look at a simple sketch to see how the photoresistor works. The photoresistor is also available in the same kit from Sunfounder. The module has three pins, with the outer pins labeled "–" and "S" as shown in Figure 4.3.

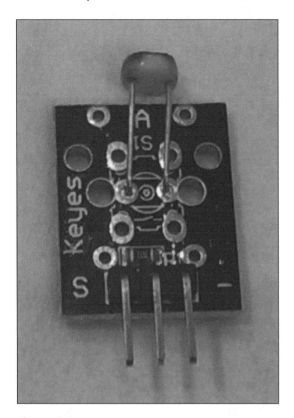

Figure 4.3
Photoresistor.

The photoresistor pin connections are shown in Table 4.2.

Table 4.2 Photoresistor Pin Connections	
Photoresistor Pin	**Arduino Pin**
−	GND
Middle	5V
S	A0

The following sketch is for using the photoresistor.

```
/* photoresistor check
*/
int sensorPin = A0;
int value = 0;

void setup ()
{
  Serial.begin (9600);
}

void loop ()
{
  value = analogRead (sensorPin);
  Serial.println (value, DEC);        // If light shines on it, value is low. If not, value is
                                      // high.

  delay (1000);
}
```

Note

We first experimented with a handheld flashlight shining on the photoresistor. With the light on, the value displayed on the monitor was about 30. With the light off, the value was about 350.

How the Code Works

A set of variables is used in this sketch. Pin A0 holds the value of the `sensorPin` and the integer variable `value` holds a value of 0.

The `loop()` function does the following:

The variable value reads the analog value of the `sensorPin`.

The `Serial.println` method prints out this value in a decimal format to the serial monitor.

A delay of one second is executed.

This sketch prints out the value of the `sensorPin` every second. If you shine a flashlight on the sensor, you can see the difference in values.

Now you can easily see how the intruder alarm works. We just put the two sketches together. In the experimental setup (Figure 4.1), we arrange the laser diode and the detector such that the laser light falls on the detector. If a burglar interrupts the laser beam, the photoresistor value becomes high, and this high value can be used to turn on an LED or generate a beep.

Sketch Combining the Laser and the Photoresistor

```
/* Intruder alarm with laser diode and
   photoresistor
*/
int sensorPin = A0;      // S pin of photoresistor connected to A0.
int value = 0;
void setup ()
{
  pinMode (8, OUTPUT); // digital output to laser
  pinMode (7, OUTPUT); // digital output to LED
  digitalWrite (8, HIGH); // laser on all the time
}

void loop ()
{
  value = analogRead(sensorPin);
  if (value > 100)
  {
   digitalWrite (7, HIGH); // LED on if beam interrupted
  }
  else
  {
    digitalWrite (7, LOW); // LED off
  }
  delay (100); // delay one-tenth of a second
}
```

How the Code Works

Two variables are created. The sensorPin (S pin of the photoresistor) is connected to A0 on the Arduino board. An integer variable called value is set to 0.

The setup() function does the following:

Digital output from the Arduino pin 8 is sent to the laser.

Digital output from pin 6 is sent to the LED.

The laser is on (HIGH) all the time, and the beam is falling on the sensor.

The loop() function does the following:

The variable value gets the output from the sensorPin. As long as the laser beam is falling on the sensor, the sensor value is low (< 100).

If the value > 100, that means no laser light is falling on the sensor, so someone has interrupted the laser beam. The LED is turned on (HIGH).

Otherwise (value < 100), it means the laser beam is falling on the sensor, there is no intruder, and the LED is turned off (LOW).

A delay of one second is used. So every second the value is checked, and based on the result the laser is turned on or off.

Activity 2: Proximity Alarm with an Ultrasonic Range Sensor

You have already used the ultrasonic range sensor in the design of the obstacle-avoiding robot in Chapter 2. The ultrasonic sensor is used to find the distance to an object within a specified range. You can set the range to a convenient distance so that if an object (or person) approaches within that distance, the system emits an audible or visual alarm.

Since you have already seen the basic sketch, which shows how the ultrasonic range sensor works, we will proceed straight to the design of the proximity alarm. Figure 4.4 shows the circuit layout, with the ultrasonic sensor and a piezo buzzer in the foreground. In this case, we included an LED and a piezo buzzer for the alarm.

Figure 4.4
Proximity alarm circuit.

Sketch for the Proximity Alarm

```
/*
Ultrasonic proximity sensor
*/

#include <NewPing.h>      // Library for the ultrasonic sensor
#define TRIGGER_PIN   5    // Arduino pin tied to trigger pin on the ultrasonic sensor.
#define ECHO_PIN      6    // Arduino pin tied to echo pin on the ultrasonic sensor.
#define MAX_DISTANCE 100  // Maximum distance we want to ping for (in centimeters). More than
                          // that will give zero ping time.
//Maximum sensor distance is rated at 400-1000cm.
NewPing sonar(TRIGGER_PIN, ECHO_PIN, MAX_DISTANCE); // NewPing setup of pins and maximum
                                                    // distance. sonar is an instance of
                                                    // type NewPing
void setup()
{
  pinMode(5, OUTPUT);
  pinMode(6, INPUT);
  pinMode(7, OUTPUT);     // Output to LED
  pinMode(8, OUTPUT);     // Output to buzzer
}

void loop()
{
    unsigned int uS = sonar.ping(); // Send ping, get ping time in microseconds (uS).
    int dist = uS / US_ROUNDTRIP_CM;   // Uses a function in the library to convert
                                       // microseconds to distance in centimeters

  if (dist < 10 || dist == 0)    // If the distance is too close or more than 100 cm
  {
    digitalWrite(7, LOW);     // LED off
  }
  else                        // If the object is within 10 cm - 100 cm
  {
    digitalWrite(7, HIGH);     // LED on
    for (int i = 0; i < 500; i ++)      // Sound the buzzer
    {
      digitalWrite(8, HIGH);    // Send ON - OFF to buzzer 500 times
      digitalWrite(8, LOW);     // This produces a buzzing sound
    }
  }
}
```

How the Code Works

The `NewPing.h` library for the ultrasonic sensor is included in the sketch. Next, a set of variables is created. The Arduino pin number 5 is mapped to the trigger pin on the ultrasonic sensor.

The Arduino pin number 6 is mapped to the echo pin on the ultrasonic sensor.

The MAX_DISTANCE that the sensor will be pinging is set to 100 cm, after which you get a zero value.

The `setup()` function does the following:

Pin 5 is used for OUTPUT. This triggers the output ultrasonic pulse.

Pin 6 is used for INPUT. This pin receives the return pulse from the object.

Pin 7 is used for output to the LED.

Pin 8 is used for output to buzzer.

The `loop()` function does the following:

The ultrasonic sensor sends a ping and returns the ping time in microseconds (uS).

US_ROUNDTRIP_CM is a function in the library that converts the microseconds to centimeters.

If the distance is too close, or <10 cm, or more than 100 cm, then the LED is Off.

Otherwise, the LED is On.

A `for` loop is used where the buzzer is toggled between HIGH and LOW for 500 microseconds. This toggle produces the sound from the buzzer.

ACTIVITY 3: TOUCH SENSOR AND ALARM

If a person touches an object, the electrical capacitance of the human body changes the electrical characteristics of a circuit close to the object. These changes can be detected and consequently an alarm can be produced. Imagine setting this up such that if someone in the house touches the cookie jar, an alarm sounds and LEDs start blinking!

A touch sensor is very easy to implement. A hookup wire is connected to one of the digital pins of the Arduino board. The other end of the wire is wrapped in aluminum foil. The foil is attached to your object of value with sticky tape over it. In our case, we attached the foil to a thick plastic sheet. If you touch the sticky tape with your finger, the digital value at the connected digital pin changes, and that can be used as a signal to light an LED. It is as simple as that. You should touch only the sticky tape, and not the aluminum foil. That is the *capacitive effect*, and that is why this sensor is called a capacitive touch sensor. Figure 4.5 shows the touch sensor layout, and Table 4.3 shows the pin connections.

Figure 4.5
Touch sensor.

Table 4.3 Touch Sensor Connections	
Touch Sensor	**Arduino Pin**
Hookup wire	7
220 Ω resistor	8 (Leads to LED)

Sketch for the Touch Sensor

```
int LED = 8; // Arduino pin connected to LED
int touchPin = 7; // Arduino pin connected to touch sensor
int val ; // define numeric variable
void setup ()
{
  pinMode (LED, OUTPUT) ; // output to LED
  pinMode (touchPin, INPUT) ; // touch sensor input
}
```

```
void loop ()
{
  val = digitalRead (touchPin) ;
  if (val == HIGH) // When the metal touch sensor detects a signal, LED flashes
    {
      digitalWrite (LED, HIGH);
    }
  else
    {
      digitalWrite (LED, LOW);
    }
}
```

How the Code Works

A set of variables is used in this sketch.

Arduino pin 8 is mapped to an LED.

Arduino pin 7 is mapped to the touch sensor.

An integer variable val is created.

The setup() function does the following:

The LED pin is set as an OUTPUT pin.

The touchPin is set as an INPUT pin.

The loop() function does the following:

The integer variable val gets the value of the touchPin.

If val is HIGH, the LED flashes.

Otherwise, val is LOW and the LED does not light up.

ACTIVITY 4: KEYBOARD AND LEDS

Just for fun, we can use the same principle of a capacitive touch sensor to make a keyboard with which you can turn on different colored LEDs, or make music, or both.

Here we connected four touch sensors to four digital pins, and used four other digital pins to light four different LEDs, as shown in Figure 4.6.

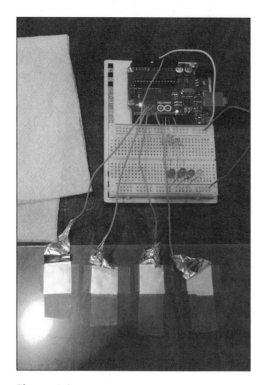

Figure 4.6
Touch sensors, pins, and LEDs.

The sketch simply extends the previous sketch to four capacitive touch sensors and four LEDs.

Sketch to Light Up LEDs with Touch

```
int ledRED = 8;// Arduino pin connected to LED
int ledBLU = 9;
int ledGRE = 10;
int ledYEL = 11;
int keyRED = 4; // Arduino pin connected to touch sensor
int keyBLU = 5;
int keyGRE = 6;
int keyYEL = 7;
int valRED, valBLU, valGRE, valYEL;// define numeric variables for values on pins
void setup ()
{
  pinMode (ledRED, OUTPUT) ;// output to LED
  pinMode (ledBLU, OUTPUT) ;
```

```
  pinMode (ledGRE, OUTPUT) ;
  pinMode (ledYEL, OUTPUT) ;

  pinMode (keyRED, INPUT) ;// touch sensor input
  pinMode (keyBLU, INPUT) ;
  pinMode (keyGRE, INPUT) ;
  pinMode (keyYEL, INPUT) ;
}
void loop ()
{
  valRED = digitalRead (keyRED) ;//
  if (valRED == HIGH) // When the metal touch sensor detects a signal, red LED glows
    {
      digitalWrite (ledRED, HIGH);
    }
  else
    {
      digitalWrite (ledRED, LOW);
    }

  valBLU = digitalRead (keyBLU) ;//
  if (valBLU == HIGH) // When the metal touch sensor detects a signal, blue LED glows
    {
      digitalWrite (ledBLU, HIGH);
    }
  else
    {
      digitalWrite (ledBLU, LOW);
    }

  valGRE = digitalRead (keyGRE) ;//
  if (valGRE == HIGH) // When the metal touch sensor detects a signal, green LED glows
    {
      digitalWrite (ledGRE, HIGH);
    }
  else
    {
      digitalWrite (ledGRE, LOW);
    }

  valYEL = digitalRead (keyYEL) ;//
  if (valYEL == HIGH) // When the metal touch sensor detects a signal, yellow LED glows
    {
      digitalWrite (ledYEL, HIGH);
    }
```

```
else
  {
    digitalWrite (ledYEL, LOW);
  }
}
```

How the Code Works

A set of variables—ledRED, ledBLU, ledGRE, and ledYEL—is created and mapped to Arduino pins 8, 9, 10, and 11, which are connected to LEDs of different colors.

Variables keyRED, keyBLU, keyGRE, and keyYEL are mapped to Arduino pins 4, 5, 6, and 7, which are connected to the touch sensor.

Variables valRED, valBLU, valGRE, and valYEL are used to take in the values from the different pins.

The setup function does the following:

The ledRED, ledBLU, ledGRE, and ledYEL pins are set for output to LED.

The keyRED, keyBLU, keyGRE, and keyYEL are used for input.

The loop function does the following:

The value from keyRED is given to the variable valRED. If this is HIGH, then the red LED lights up; otherwise it is LOW and the LED does not light up.

This same sequence is repeated for keyGRE, keyBLU, and keyYEL.

CONCLUSION

In this chapter, you learned how to use different types of sensors to design alarms for different situations. You used a diode laser and a photoresistor to set up an intruder alarm. Then you used the familiar ultrasonic range finder to design a proximity alarm. The third sensor you designed and used was a capacitive touch sensor to detect when someone touches a precious object, such as a cookie jar! Finally, as a fun project, you used the same capacitive touch sensors to design a simple keyboard with which you can light up LEDs or play music.

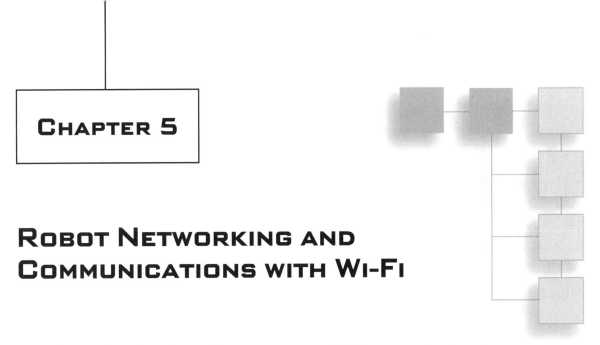

CHAPTER 5

ROBOT NETWORKING AND COMMUNICATIONS WITH WI-FI

This chapter describes the tools necessary to establish communications between a computer and a robot using Wi-Fi, and setting up a server for the robot.

CHAPTER OBJECTIVES

- Pair your Arduino with a Wi-Fi shield
- Learn the uses of the Wi-Fi library
- Connect to wireless networks
- Program a basic server application
- Send messages to/from your computer using Wi-Fi

INTRODUCTION

Wireless communication has been around since the late 1890s, when inventors such as Alexander Graham Bell, Guglielmo Marconi, and others discovered that electromagnetic waves were capable of sending information through the air. Since then, radio waves have evolved to carry not only analog audio and video content through radio and antenna television, but also digital data at ever-increasing speeds and precision.

Cellular telephones were the next evolutionary jump, offering not only widespread receiving capabilities, but also individual lines of analog communication. By 1987, one million people had subscribed to cellular telephone plans, and by 1990, there were five million cellular users. Today, there are over six billion cellular users!

In 1999, the term *Wi-Fi* was coined to describe the IEEE 802.11 specifications, paving the way for mobile computers and faster Internet connections. Wi-Fi (IEEE 802.11b) usage began to spread with mobile phones and portable computers, providing wireless Internet services in mobile applications as well as local network access in fixed locations. The standard has expanded, and devices have been built for various versions of Wi-Fi (802.11), including b, g, n and ac. Each version increases the speed of the last, jumping from 11 Mbits/s (megabits per second) in 802.11b to version 802.11ac, which can reach multiple Gbits/s (gigabits per second)!

How does Wi-Fi work? You probably know the basics: A wireless router connects to the Internet via a modem and broadcasts a wireless signal. Clients can then connect to this wireless signal and transfer data between connected devices and devices available on the Internet via the modem. If a modem is not available, internal network traffic can continue to work over a LAN (Local Area Network).

To make this magical connection happen, there are numerous layers in place to get your data to your screen quickly and accurately. At the physical layer are a wireless transmitter and receiver, which emit electromagnetic waves that can be modulated to transfer information. Each device has a fixed unique identifier (MAC address) as well as an assigned virtual address (IP address), which are used to designate where information is sent and received. Each packet of data is sent with a heading that denotes (among other things) who sent the data, where it should go, and what protocol it is following. Software on both connecting devices converts these packets into meaningful data and displays it or processes it appropriately.

While the Arduino Uno you are using is not equipped with onboard Wi-Fi antennas, an external shield provides this functionality. Shown in Figure 5.1, this shield contains an antenna and communicates with the Arduino via serial connections. Note that some Wi-Fi shields must be updated to the most recent firmware to be used. If you experience problems starting this project, please visit our companion website to view instructions on upgrading your shield's firmware.

Figure 5.1
An Arduino Wi-Fi shield.

Materials Required

- Standard Arduino robot used in previous chapters
- Arduino Wi-Fi shield
- USB mini cable
- Ultrasonic range finder
- Wi-Fi network

Part 1: Installing the Wi-Fi Sensor and Connecting to Your Network

We will begin by installing the shield onto the Arduino by placing it on top. The pins on the shield will fall perfectly into the Arduino. Push firmly, but not too hard, and make sure all the pins go into their sockets.

1. Start by separating your Arduino from your robot chassis.
2. Slowly place the Wi-Fi shield onto the Arduino, pushing it in firmly.
3. Plug the Arduino Uno into your computer using your USB cable (no other power source is needed).

To test the Wi-Fi sensor, we will write a basic sketch that will list all available networks, which we will view over the serial-to-USB interface.

4. Open the Arduino application on your computer and create a new sketch.

5. At the top of the sketch, we will import two files, SPI and WiFi, using the following lines:

```
#include<SPI.h>
#include<WiFi.h>
```

6. In the setup loop, we will start the serial output and wait until a connection occurs by entering

```
Serial.begin(9600);
while(!Serial);
```

7. We will now create a new void function called printNetworks() with the following signature:

```
void printNetworks()
```

8. Start the function by printing a status message:

```
Serial.println("Scanning Networks");
```

9. To get a list of Wi-Fi networks, use the function WiFi.scanNetworks() like so:

```
byte num = WiFi.scanNetworks();
```

This returns the number of available Wi-Fi networks, which can be used to get more data from the Wi-Fi library.

10. Create a for loop to cycle through all possible network IDs by counting from 0 until the number of networks we just received from the previous command:

```
for(int network =0; network<num; network++)
```

Within the loop, we will print data about each network. The primary three pieces of information are the network name, network strength, and the encryption protocol. We will use the following commands:

WiFi.SSID(network) Returns the Network SSID or name.

WiFi.RSSI(network) Returns the Network's strength in dBm.

WiFi.encryptionType(network) Returns what type of security this network uses (WEP, WPA, etc.).

11. Print the network data (via serial), and then return a new line for the next network using the following block of code:

```
Serial.print(network);
Serial.print(") ");
Serial.print(WiFi.SSID(network));
Serial.print("   Strength: ");
Serial.print(WiFi.RSSI(network));
Serial.print(" dBm   Security: ");
Serial.println(WiFi.encryptionType(network));
```

12. Finally, call the printNetworks() function in your loop(), with a 10-second delay between calls. When complete, your code should look like the code below.

```
#include <SPI.h>
#include <WiFi.h>

void setup() {
  // Start Serial, Waiting For Connection:
  Serial.begin(9600);
  while(!Serial) ;

}

void loop() {
 delay(10000);
 printNetworks();

}

void printNetworks() {
  // scan networks:
  Serial.println("Scanning Networks");
  byte num = WiFi.scanNetworks();

  // print the network number and name for each network found:
  for (int network = 0; network<num; network++) {
    Serial.print(network);
    Serial.print(") ");
    Serial.print(WiFi.SSID(network));
    Serial.print("   Strength: ");
    Serial.print(WiFi.RSSI(network));
    Serial.print(" dBm   Security: ");
    Serial.println(WiFi.encryptionType(network));
  }
}
```

13. Compile and upload your sketch to the Uno (updating the board and port if necessary). Wait for the IDE to report that it is "Done Uploading."

14. Open the serial monitor under Tools > Serial Monitor. If text does not print soon, you may need to reset your Arduino. Ensure that it is set to 9600 baud.

15. Your serial monitor will display the available networks, refreshing every 10 seconds, as shown in Figure 5.2. If nothing appears, or an error occurs, you may need to update your shield's firmware. Instructions for doing so can be found on the companion website (www.cengageptr.com/downloads).

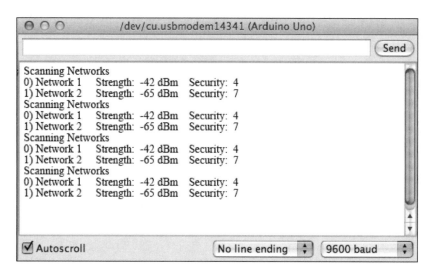

Figure 5.2
Serial monitor output.
Source: Arduino Group.

Notice in the serial monitor that the networks are listed numerically, displaying the network name (SSID) as you see on your computer. The Strength number (usually something between 0 and −100) indicates how strong your wireless signal is, while the Security number denotes the type of network detected. The Security numbers correspond with the security type as shown in Table 5.1.

Table 5.1   Security IDs

ID	Type
5	WEP
2	WPA
4	WPA2
7	None
8	Auto

Most networks utilize 7 (no security) or 4 (WPA), allowing easy connections. We will perform both connections now.

Connecting to an Open Network

To connect to a network, we will modify our existing sketch by creating a function and calling it in the setup() function.

1. Create a new function with the following header in your existing sketch:

```
void connectOpenNetwork(char ssid[])
```

The ssid array will provide the name of the network, which will be entered when the function is called.

2. Create a global variable at the very beginning of the sketch:

```
int status = WL_IDLE_STATUS;
```

WL_IDLE_STATUS is a status that indicates that the Wi-Fi is doing nothing.

3. In the connectOpenNetwork() function, print a status message declaring which network you are connecting to.

```
Serial.print("Connecting To ");
Serial.println(ssid);
```

4. Connect to the network using the WiFi.begin(id) function as follows:

```
WiFi.begin(ssid);
```

5. This `begin` function returns a status indicator, so we will modify the line to assign it to status.

```
status = WiFi.begin(ssid);
```

6. To wait until the network is connected, use a `while` loop, checking until status is equal to WL_CONNECTED.

```
while(status!=WL_CONNECTED);
```

7. Print a confirmation message to show you are connected, and to print the connected IP address.

```
Serial.print("Connected to: ");
Serial.println(WiFi.localIP());
```

The completed function is as follows.

```
void connectOpenNetwork(char ssid[])
 {
    Serial.print("Connecting To ");
    Serial.println(ssid);
    status = WiFi.begin(ssid);
    while(status!= WL_CONNECTED);
      Serial.print("Connected to: ");
      Serial.println(WiFi.localIP());


 }
```

Connecting to a Closed Network

To connect to a closed (password-protected) network, you will need to write a second function: `connectClosedNetwork()`.

1. Start by copying and pasting the entire `connectOpenNetwork()` function you just created.

2. Rename the copied function to `connectClosedNetwork()`.

3. Add a second argument, `char password[]` to pass a password.

4. Modify the `WiFi.begin(ssid)` line by adding the second argument:

```
status = WiFi.begin(ssid,password);
```

5. Your completed function is as follows.

```
void connectClosedNetwork(char ssid[], char password[])
 {
    Serial.print("Connecting To ");
    Serial.println(ssid);
    status = WiFi.begin(ssid,password);
    while(status!= WL_CONNECTED);
   Serial.print("Connected to: ");
   Serial.println(WiFi.localIP());

 }
```

6. You can now test these functions by calling them in the `setup()` function. If your network is unsecured, use

```
connectOpenNetwork("Network 1");
```

7. Replace Network 1 with the name of your network. If your network has a password, use

```
connectClosedNetwork("Network 1", "myPassword");
```

8. Replace Network 1 with your network name and myPassword with your password. Compile and run your program. Open the serial monitor to confirm that the device connects. If it does not connect, make sure your network name and password are correct.

PART 2: CREATING A TELNET SERVER

Now that you are connected to the network, you want to transfer data. Unfortunately, because networks are so flexible with various devices, you have to use the concept of clients and servers to communicate between your Arduino and your computer. Telnet is an easy-to-use protocol that defines how the server and client communicate, acting as a plain-text communication between a Telnet client, such as PuTTY or Terminal, and a server on your Uno.

To create this connection, you will need two special pieces of software that can be downloaded from our companion website (www.cengageptr.com/downloads): Arduino IDE 1.0.2 and PuTTY.

If you are using Mac OSX or Linux, you do not need to download PuTTY, as Telnet is pre-installed in your Terminal application. Due to bugs in version 1.0.5 of Arduino IDE, a version prior to 1.0.5 is needed; otherwise, connections are refused by the Arduino. We will be modifying the sketch you have been using in this project, so if you need to switch IDEs, copy and paste the code into a new sketch using the correct IDE.

1. Create the server by creating the following global variable at the top of the sketch:

```
WiFiServer server(23);
```

In this case, 23 is the port number. 23 is the standard Telnet port.

2. Create a second global variable to keep track of your client's connection status:

```
boolean clientConnected = false;
```

3. Add the following line to the end of your setup function to start the server:

```
server.begin();
```

Your setup function will now look like the following code. The connect function will be based on your network configuration as described in the previous section.

```
int status = WL_IDLE_STATUS;
WiFiServer server(23);
boolean clientConnected = false;

void setup() {
    // Start Serial, Waiting For Connection:
    Serial.begin(9600);
    while(!Serial) ;
    connectOpenNetwork("Network");
    server.begin();
}
```

4. Go to your `loop()` function. Connect to an incoming client by adding the following line:

```
WiFiClient client = server.available();
```

5. We will now interact with the client by using the following block of code inside the `loop` function. It is commented to explain each piece.

```
void loop() {
    WiFiClient client = server.available(); // Create connection to client.
    if(client) // Make sure the client is active/
    {
        if(!clientConnected) // If the client is not already connected, we'll print a
message.
```

```
    {
        client.flush(); // Clears the communication channel.
         Serial.println("New Client"); // Alerts us via serial that a new connection
                                        // is here.
        client.println("Hi!");   // Sends a message to the client.
          clientConnected=true; // Establishes that a client is connected, so we
                                // don't do this again.
    }
    if(client.available()>0) // Checks if the client has sent any data.
    {
        char temp = client.read(); // Read the character sent by the client and save
                                   // it.
        server.write(temp); // Send the character back to the client.
    }
  }
}
```

6. Compile your code onto your Arduino.

Now that your Arduino is ready to act as a server, we will prepare your computer to send and receive data from it.

1. Make sure you are connected to the same network as your Arduino. The SSIDs should be the same.

2. Connect to your Arduino from the serial monitor, confirm that it connects, and note the IP address it prints.

For Microsoft Windows Users

1. Open your command prompt from the Start menu.

2. Ensure that you can reach your Arduino by running the "ping" command as follows, where 192.168.1.5 is the IP address printed by your Arduino:

```
ping 192.168.1.5
```

Press Enter to run the command. You should see multiple lines that begin with "Reply from 192.168.1.5: bytes=32 ...." This confirms the robot can be reached. Press Ctrl + C to end the command. If the ping command does not return with replies, reset your Arduino and ensure that your computer is on the same network.

3. Download PuTTY from the companion website (www.cengageptr.com/downloads). Open the application.

4. Enter the IP address printed by the Arduino in the Host Name box. Change the Connection Type to Telnet. Press Connect. A black box will appear with a message from your Arduino (see the next step; we printed "Hi!").

5. Type "Hi!" into the box. Your characters are sent to the Arduino, which echoes them back, printing them to your screen.

For Mac/Linux Users

1. Open Terminal from your Applications folder.

2. Type the ping command as follows, where 192.168.1.5 is the IP address printed by your Arduino:

```
ping 192.168.1.5
```

Press Enter to run the command.

You should see multiple lines that begin with "Reply from 192.168.1.5: bytes=32 ...." This confirms the robot can be reached. Press Control + C to end the command. If the ping command does not return with replies, reset your Arduino and ensure that your computer is on the same network.

3. To Telnet into the Arduino, run the command

```
telnet 192.168.1.5
```

where 192.168.1.5 is the IP address returned by your robot. Press Enter to run the command. A message should print to the screen with the "Hi" message you created on the Arduino upon connection (see the next step).

4. Type any message into the terminal. Your text characters are sent to the Arduino, which echoes them back, printing them to your screen.

Now that you can send data back and forth between your computer and the Arduino, we will set up a command to retrieve data from your robot from the computer, via Wi-Fi. The plan is to use the characters sent from the computer as op-codes, which signify what the robot should do. We will use the letter "u" as the command to retrieve the sensor value of an ultrasonic sensor. Please review Chapter 2, "Build Your Own Robot Sweeper," if you do not remember how ultrasonic sensors work or how to implement them in Arduino.

1. Wire an ultrasonic sensor to your robot. For this example, we will plug the trigger pin into port 7 and the echo pin into port 6. Instantiate the sensor with the `NewPing` function.

2. In the `loop` function, change the initial client message to send "Press u to retrieve the ultrasonic sensor value."

3. In the last `if` statement where we read the character from the client and return it, create a new `if` statement to replace the `server.write(temp)` line.

```
if(client.available() >0) // Checks if the client has sent any data
    {
        char temp = client.read(); // Read the character sent by the client and save it.
        if(temp == 'u')
        {
            unsigned int uS = sonar.ping(); // Get the ping time
            server.write(uS/US_ROUNDTRIP_CM); // Send the ping distance
        }
    }
```

4. Re-upload your Arduino code and check the serial monitor to get the Arduino's IP address.

5. Connect via Telnet as described above. Type the letter "u" and press Enter. The Arduino will respond with its current distance from the ping sensor, typically the distance from the wall.

THE COMPLETE SKETCH

```
#include <SPI.h>
#include <WiFi.h>
#include <NewPing.h>
int status = WL_IDLE_STATUS;
WiFiServer server(23);
boolean clientConnected = false;
NewPing sonar(6, 7, 200);
void setup() {
  // Start Serial, Waiting For Connection:
  Serial.begin(9600);
  while(!Serial) ;
  connectOpenNetwork("Network");
  server.begin();
}

void loop() {

   WiFiClient client = server.available(); // Create connection to client.
   if(client) // Make sure the client is active/
```

```
{
    if(!clientConnected) // If the client is not already connected, we'll print a
                         // message.
    {
       client.flush(); // Clears the communication channel.
        Serial.println("New Client"); // Alerts us via serial that a new connection is
                              // here.
       client.println("Hi!");   // Sends a message to the client.
        clientConnected=true; // Establishes that there is a client connected, so we
                             // don't do this again.
    }
    if(client.available() >0) // Checks if the client has sent any data.
    {
       char temp = client.read(); // Read the character sent by the client and save it.
       if(temp == 'u')
       {
       unsigned int uS = sonar.ping(); // Get the ping time
       server.write(uS/US_ROUNDTRIP_CM); // Send the ping distance
       }
    }
  }
}

void printNetworks() {
  // scan networks:
  Serial.println("Scanning Networks");
  byte num = WiFi.scanNetworks();

  // print the network number and name for each network found:
  for (int network = 0; network<num; network++) {
    Serial.print(network);
    Serial.print(") ");
    Serial.print(WiFi.SSID(network));
    Serial.print("  Strength: ");
    Serial.print(WiFi.RSSI(network));
    Serial.print(" dBm    Security: ");
    Serial.println(WiFi.encryptionType(network));
  }


}
```

```
void connectOpenNetwork(char ssid[])
 {

    Serial.print("Connecting To ");
    Serial.println(ssid);
    status = WiFi.begin(ssid);
    while(status!= WL_CONNECTED);
      Serial.print("CONNECTED!");
          Serial.print(WiFi.localIP());


  }

void connectClosedNetwork(char ssid[], char password[])
 {

    Serial.print("Connecting To ");
    Serial.println(ssid);
    status = WiFi.begin(ssid,password);
    while(status!= WL_CONNECTED);
     Serial.print("CONNECTED!");
     Serial.print(WiFi.localIP());

  }

  void printNetworks() {
  // scan networks:
  Serial.println("Scanning Networks");
  byte num = WiFi.scanNetworks();

  // print the network number and name for each network found:
  for (int network = 0; network<num; network++) {
    Serial.print(network);
    Serial.print(") ");
    Serial.print(WiFi.SSID(network));
    Serial.print(" Strength: ");
    Serial.print(WiFi.RSSI(network));
    Serial.print(" dBm    Security: ");
    Serial.println(WiFi.encryptionType(network));
  }
}
```

CONCLUSION

You have successfully added Wi-Fi functionality to your Arduino via the Wi-Fi shield. You have seen how Wi-Fi networks can be searched and printed through the Wi-Fi library, as well as how to connect to a network. You then created a basic server application which not only responded to all input, but also gathered data using the robot's sensors and returned the data via Wi-Fi! Using this knowledge, you can transform all of your previous applications into remote controlled robots, gathering data live without having to save and transfer data. Wireless communication via Wi-Fi has forever changed how we use mobile devices, allowing instantaneous transfer of huge amounts of data, and now, your Arduino has joined the mobile revolution.

CHAPTER 6

ROBOT MEDICAL ASSISTANT

In this chapter, you will design a robot medical assistant that will remind a patient to take medication at specified times.

CHAPTER OBJECTIVES

- Design a robot medical assistant—a prescription reminder—using an Arduino board
- Use the `Timer()` function
- Use the pitches.h library to access different types of sound files
- Use the LCD panel to display messages to the user

INTRODUCTION

What is a robot medical assistant? That can be many things, from a simple pill reminder (which we will create in this chapter) to a robot that records your vital signs at regular intervals, logs the data, reminds you of the medications you need to take at the appropriate time(s) of day, counts or measures dosages to ensure accuracy, reminds you to refill your medications, and many other things. It could even be programmed to caution you if any of your vital signs exceed prescribed limits, so you would know when to go see your doctor.

In this chapter, you will design and build a simple version of a prescription reminder using an Arduino board. At specified intervals, the robot medical assistant prompts you to take a specified medicine by producing a musical note and displaying the appropriate message on an LCD display.

MATERIALS REQUIRED

■ Arduino board

■ LCD display

■ Push button

■ Trimmer potentiometers (2)

■ Speaker

■ Solderless breadboard and hookup wires

PROGRAM COMPONENTS AND CONNECTIONS

The components of our program include:

■ A timer to keep track of the times when the reminder is necessary

■ A set of musical notes to be played at the appropriate times

■ A set of messages to be displayed on the LCD panel, each at the appropriate time

The connections to be made are shown in Tables 6.1 to 6.5.

Table 6.1 LCD Panel Connections

LCD Pin	Arduino Pin
1	GND
2	5V
3	Tap on Pot 1
4	7
5	GND
6	8
7	Blank
8	Blank

9	Blank
10	Blank
11	9
12	4
13	5
14	6
15	5V
16	GND

The Arduino pins 7, 8, 9, and 4, 5, 6 must be specified in an LCD panel initialization in the sketch.

Table 6.2 Potentiometer 1 Connections

Pot 1	Connected To
(+Vcc)	5V
Middle	3 on LCD
(-Vcc)	GND

Table 6.3 Potentiometer 2 Connections

Pot 2	Connected To
(+Vcc)	5V
Middle	Arduino pin A0
(-Vcc)	GND

Table 6.4 Push Button Connections

Button	Connected To
Pin 1	5V
Pin2	Arduino pin 3
	&1kΩ to GND

Table 6.5 Speaker Connections

Speaker	Connected To
(+Vcc)	Arduino pin 2
(-Vcc)	GND

The connections for the LCD panel will be used again in the next chapter. The screw on the potentiometer (Pot 1) is used to change the contrast on the LCD panel. Pot 2 is used to set the delay in the Timer() function. The push button is used to reset Arduino pin 3 to HIGH, in readiness for the Timer() function to be used.

The critical feature in this design is the Timer() function. The Timer() function is used to set the interval between dosages. The push button and the potentiometer (Pot 2) in the circuit are the critical hardware components used to set the time delay, which can be set to any value up to 24 hours.

Figure 6.1 shows the functionality of the push button.

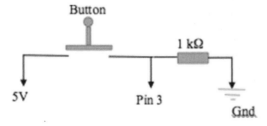

Figure 6.1
Push button.

If the button is pressed and released, it sets Arduino pin 3 to 5V (HIGH). The Timer() function sets pin 3 to 0V (LOW) and then reads the value of the voltage on Arduino pin A0 to set the time. The 1 kΩ resistor is to prevent the power supply (5V) from being shorted to ground.

Figure 6.2 is a schematic diagram of the potentiometer (Pot 2).

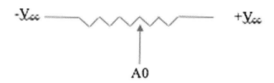

AO

Figure 6.2
Potentiometer.

The potentiometer has three pins. The connections are made as shown in the table: the $-V_{cc}$ pin of the pot is connected to ground, and the $+V_{cc}$ pin is connected to +5V. The middle pin is connected to the A0 pin of the Arduino board. The middle pin is a slider, and as it slides from one end of the potentiometer to the other, the voltage at the middle pin changes from 0V to 5V continuously. This voltage is read by the Arduino and is converted to a proportional time delay.

Figure 6.3 shows the complete setup of the robot medical assistant.

Figure 6.3
Robot medical assistant setup.

PILL REMINDER SKETCH

The following sketch is the code required to execute the medication reminder project.

```
// include the library code for the LCD display:
#include <LiquidCrystal.h>
// Library to set musical notes for the Alarm() function
#include "pitches.h"

// initialize the display with the numbers of the interface pins
LiquidCrystal lcd(7, 8, 9, 4, 5, 6);    // Initializing digital pins for LCD panel
int melody[] = {NOTE_C4, NOTE_G3};      // notes in the Alarm melody
int noteDurations[] = {4, 8};           // note durations: 4 = quarter note, 8 = eighth note, etc.
const int buttonPin = 3;                // the number of the push button pin to set the Timer
int buttonState = 0;                    // variable for reading the push button status
const int analogInPin = A0;             // Analog input pin that the potentiometer (Pot 2) is
                                        // attached to
int sensorValue = 0;                    // value read from A0
unsigned long time, hours, minutes;     // variables to hold values for delay time

void setup()
{
// set up the LCD's number of columns and rows:
  lcd.begin(16, 2);
  // Print a message to the LCD.
  lcd.print("Initializing...");
  delay(5000);
  Timer();    // Call Timer() function, wait for push button
  Display1();       // Ask patient to relax before taking medicine
}

void loop()          // Main loop
{
  delay(time);       // As set with the Timer() function
  AlarmDisplay(); // Alert patient "Time to take your medicine"
  Alarm();           // Sound musical notes
  Display2();        // "Take the red headache pill"
  delay(time);       // Set with Timer()
  AlarmDisplay(); // Alert patient "It is time for the next dose"
  Alarm();           // Sound musical notes
  Display3();        // "Take the little blue pill"
}

void Display1()                      // Print a message to the LCD.
{
  lcd.setCursor(0, 0);
  lcd.print("Relax before the ");
```

```
    lcd.setCursor(0, 1);
    lcd.print("first dosage.    ");
}

void Display2()                    // Print a message to the LCD.
{
    lcd.setCursor(0, 0);
    lcd.print("Take the red      ");
    lcd.setCursor(0, 1);
    lcd.print("headache pill.    ");
}

void Display3()                    // Print a message to the LCD.
{
    lcd.setCursor(0, 0);
    lcd.print("Take the little   ");
    lcd.setCursor(0, 1);
    lcd.print("blue pill.        ");
}

void AlarmDisplay()                // Print a message to the LCD.
{
    lcd.setCursor(0, 0);
    lcd.print("Time to take      ");
    lcd.setCursor(0, 1);
    lcd.print("your medicine.    ");
}

void Alarm()
{
    buttonState = LOW;             // Resets the state of the button to low.
    while(buttonState == LOW)      // check if the push button is pressed.
    {
        for (int thisNote = 0; thisNote < 2; thisNote++)   // iterate over the notes of the
                                                           // melody
        {
            int noteDuration = 1000/noteDurations[thisNote];    // Fix the duration of the note.
            tone(2, melody[thisNote],noteDuration);     //Produce the note at pin 2.
            int pauseBetweenNotes = noteDuration * 1.30;
            delay(pauseBetweenNotes);
            noTone(2);      // stop the note playing
        }
        buttonState = digitalRead(buttonPin);     // read the state of the push button value
    }
}
```

```
void Timer()
{
  buttonState = LOW;            // Resets the state of the button to low.
  while(buttonState == LOW)    // check if the push button is pressed.
  {
    sensorValue = analogRead(analogInPin);    // Reads the analog value from the
                                              // potentiometer at A0, "from 0 - 1023".
    hours = sensorValue/42.625;               // Converts value into hours. "Max 24 hour"
    lcd.setCursor(0, 0);
    lcd.print("Timer set to    ");
    lcd.setCursor(0, 1);
    lcd.print(hours);                         // Displays current value setting of hours.
    lcd.print(" hours    ");
    buttonState = digitalRead(buttonPin);     // read the state of the push button value.
  }
// User pushes button again to set minutes
  delay(1000);
  buttonState = LOW;
  while(buttonState == LOW)    // check if the push button is pressed.
  {
    sensorValue = analogRead(analogInPin);    // Reads the analog value from the
                                              // potentiometer at A0, "from 0 - 1023".
    minutes = sensorValue/17.05;              // Converts value into minutes.
"Max 60 minutes"
    lcd.setCursor(0, 0);
    lcd.print("Timer set to    ");
    lcd.setCursor(0, 1);
    lcd.print(minutes);                       // Displays current value setting of
                                              // minutes.
    lcd.print(" minutes");
    buttonState = digitalRead(buttonPin);     // read the state of the push button value
  }
  time = hours*3600000 + minutes*60000;       // Sets time delay between dosages.
}
```

How the Code Works

Establish variables and pin declarations.

Include the libraries required for the LCD display and for setting the musical notes for the `Alarm()` function.

Initialize the display with the numbers used for the interface pins. Pins 4, 5, 6, 7, 8, and 9 will be used as the digital pins for the LCD.

NOTE_C4 and NOTE_G3 are the notes from the notes library that will be used to play the music in the Alarm() function. The duration of the notes is also set using 4 for a quarter note and 8 for an eighth note.

Pin number 3 is the push button pin set to the timer.

A variable buttonState for reading the push button status is set to 0.

A0 is the Analog input pin to which the potentiometer is attached. The variable sensorValue is used to read the value from A0.

The variables time, hours, and minutes are used to hold values for the delay time between doses.

Void setup() does the following:

Sets up the number of rows (16) and columns (2) on the LCD panel.

Prints the message "Initializing" on the display.

Waits for 5 seconds.

Calls the Timer() function.

Calls the Display1() function.

Void loop() does the following:

Activates the delay.

Calls the AlarmDisplay() function that alerts the patient via the LCD.

Calls the Alarm() function that plays the musical notes.

Calls the Display2() function that alerts the patient to take the red headache pill.

Calls the Alarm() function that plays the musical notes.

Calls the Display3() function that alerts the patient to take the little blue pill.

Void Display1() does the following:

Sets the cursor at position 0,0 on the LCD panel.

Prints the line "Relax before the "

Moves the cursor to 0,1.

Prints "First dosage"

Void Display2() does the following:

Sets the cursor at position 0,0 on the LCD panel.

Prints the line "Take the red "

Moves the cursor to 0,1.

Prints "headache pill"

Void Display3() does the following:

Sets the cursor at position 0,0 on the LCD panel.

Prints the line "Take the little "

Moves the cursor to 0,1.

Prints "blue pill"

Void AlarmDisplay() does the following:

Sets the cursor at position 0,0 on the LCD panel.

Prints the line "Time to take "

Moves the cursor to 0,1.

Prints "your medicine"

Void Alarm() does the following:

Resets the buttonState to LOW.

While buttonState is LOW, iterates over the notes of the melody, sets a duration for each note, then plays the note for the set duration, pauses between the notes, and reads the state of the push button's value. Each note is a square wave of the appropriate frequency generated at pin 2, which is connected to the speaker. The frequency for each note is set by the library "pitches.h".

Void Timer() does the following:

Resets the state of the button to LOW.

While buttonState is LOW, reads the analog value from the potentiometer at A0.

Converts the values into hours, sets the cursor at 0,0 on the LCD panel, prints "Timer set to" on the LCD panel, moves the cursor to position 0,1, prints the value of hours and the word "hours", and again reads the state of the push button's value.

The user pushes the button again to set the minutes. The function sets `buttonState` = LOW and while the `buttonState` is LOW, reads the analog value from the potentiometer at A0 and converts the value to minutes.

This value is then printed on the LCD panel, as in the previous code block where the hours were set and printed.

The `buttonState` of the push button is read again.

The delay time is finally set by assigning the value of the hours and minutes set earlier and converting them to milliseconds. This is accomplished by the line:

Time = hours * 3600000 + minutes * 60000;

CONCLUSION

In this chapter, we designed a prescription medication reminder. The user can set the dosage times, and the system displays predetermined messages with alarms at the prescribed times. Note that the same technique can be used in many different applications—for example, to make a task list for the day.

CHAPTER 7

DATA LOGGER

In this chapter, you will learn how temperature data can be generated using a temperature sensor. You will also learn how to display the temperature data on an LCD panel, how to log data to a memory card, and how to read data from a memory card.

CHAPTER OBJECTIVES

- Use a temperature sensor to read temperature
- Display the temperature on an LCD panel
- Prepare (initialize) an SD card reader for writing or reading data
- Log data to an SD card
- Read data from an SD card

INTRODUCTION

Data logging and retrieval are essential tasks in many different applications, such as weather monitoring, industrial process monitoring, patient health monitoring, etc. In all such applications, data is generated by reading sensors, such as temperature sensors, pressure sensors, humidity sensors, and so on. The data is logged, or saved, as it is received, and then retrieved and processed. In some cases, data is processed live.

This chapter introduces you to the technique of data logging and retrieval, using an SD card shield for the Arduino board. The SD card shield can store (log) data in an SD memory card and read data from it.

We will use a specific temperature sensor to generate data, although any sensor that provides any kind of data can be used to learn about data logging.

This chapter is organized in two parts.

Part 1 describes the use of a temperature sensor to measure temperature and an LCD panel to display the temperature.

Part 2 describes the activities involved in data logging. These are:

- Activity 1: Initializing the SD Card Reader
- Activity 2: Writing Data to an SD Card
- Activity 3: Reading Data from a File
- Activity 4: Logging Temperature Data

MATERIALS REQUIRED

- TMP36 temperature sensor
- 16 × 2 LCD panel and 10 kΩ potentiometer
- Arduino board and solderless breadboard
- Stackable SD card reader
- SD card (not SDHC card)

PART 1: MEASURING AND DISPLAYING AMBIENT TEMPERATURE

The TMP36 is a temperature sensor available from Analog Devices (www.analog.com). The data sheet is at the following website:

http://www.analog.com/static/imported-files/data_sheets/TMP35_36 _37.pdf

The sensor comes in a couple different packages. We used the TO-92 package with three pins as shown in Figure 7.1.

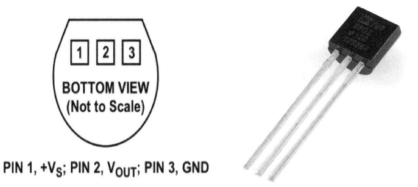

PIN 1, +V$_S$; PIN 2, V$_{OUT}$; PIN 3, GND

Figure 7.1
Pinout diagram and physical layout of TMP36.

Hardware Connections

The pin connections between the TMP36 and the Arduino board are as listed in Table 7.1.

Table 7.1 TMP36 Board Pin Connections

TMP36	Arduino
+V$_s$ (1)	5V
OUT (2)	A0
GND (3)	GND

The connections are shown in Figure 7.2. The OUT pin of the TMP36 gives an integer value between 0 and 1023, which is related to the temperature of the sensor. In the LCDTemp sketch that follows, the raw value is converted to Celsius temperature and then displayed. The display can be done on the serial monitor, but we also added an LCD panel, just for fun, to show the temperature.

Figure 7.2
TMP36 board pin connections.

The LCD panel we used is a 16 × 2-character panel available through Amazon or Ada-fruit as listed in the materials required for this chapter. You will need to do some wiring for this, but clear instructions are available at http://learn.adafruit.com/character-lcds. Follow the instructions on the website to connect the LCD panel on the breadboard, and then make the changes as indicated in the following paragraph.

The pin numbers used in the Adafruit instructions are altered in our example below. For our example, Arduino pins 10–13 are needed for data logging and cannot be used for the LCD panel. So, we used pins 4, 5, and 6 instead. The detailed pin connections for the LCD panel are listed in Table 7.2.

Table 7.2   LCD Panel Pin Connections

LCD	Arduino
1	GND
2	5V
3	Tap on Pot
4 RS	7
5	GND

6 EN	8
7	Blank
8	Blank
9	Blank
10	Blank
11 DB4	9
12 DB5	4
13 DB6	5
14 DB7	6
15	5V
16	GND

Carefully make the changes in the wiring as indicated in the Table 7.2. The connections are shown in Figure 7.3.

Figure 7.3
Completed pin connections for the LCD panel.

The Arduino pins 7, 8, 9, and 4, 5, 6 have to be specified in an LCD panel initialization in the sketch, and we therefore listed them again in Table 7.3.

Table 7.3   Arduino to LCD Pin Connections

LCD Pin	RS	EN	DB4	DB5	DB6	DB7
Arduino Pin	7	8	9	4	5	6

Writing the Sketch

In the following LCDTemp sketch, the temperature is read from the TMP36 and displayed on the LCD panel. Open the Arduino IDE and type the following sketch.

```
// Testing the analog temperature sensor and the LCD Panel

// include the library code for the LCD display:
#include <LiquidCrystal.h>

//The following command is necessary for the LCD display library
//to recognize the Arduino pins connected to the LCD panel.
LiquidCrystal lcd(7, 8, 9, 4, 5, 6);

float tmp;   //Variable to receive temperature value

void setup()
{
  Serial.begin(9600);
  // set up the LCD's number of columns and rows:
  lcd.begin(16, 2);

  // Print a message to the LCD.
  lcd.print("Temperature: ");
}

void loop()
{
  Serial.println (analogRead(A0)); // Print the raw data from the sensor on the
                              //serial monitor
  tmp = tempC();
  // Call the function to read the sensor, calculate the temperature,
  //and assign the value to our variable tmp.
  Serial.print(tmp);
  Serial.println(" C");
  lcd.setCursor(0, 1);   // Move the LCD panel cursor to the second line
```

```
  lcd.print(tmp);
  lcd.print(" C");

  delay(1000);      // Wait for 1 second and loop
}
float tempC( )
{
//Read the output from temperature sensor, calculate the temperature,
//and return the temperature.
  float tempVal, temp;
  // variables to hold intermediate values used in the calculation of temperature
  tempVal = analogRead(A0); // This is the raw output from the TMP36 sensor
  temp = ((tempVal*5000/1024)-500)/10; // Raw data converted to Celsius degrees
  return temp;
}
```

How the Code Works

Include the library code for the LCD display.

Initialize the LCD panel with the correct pin assignments necessary for the LCD display library to recognize the Arduino pins connected to the LCD panel.

A float variable temp is created to hold the value of the temperature in Celsius.

The setup() function does the following:

Opens the serial port with a baud rate of 9600.

Sets up the LCD panel with 16 columns and 2 rows.

Prints the word "Temperature " on the LCD panel.

The loop() function does the following:

Prints the value from Arduino pin A0 on the serial monitor.

Assigns the value from the function tempC to the variable temp.

Prints the value of temp on the serial monitor.

Prints the string " C" on the serial monitor.

Sets the cursor at column 0 and row 1 on the LCD panel.

Prints the value of temp on the LCD panel.

Prints the string " C" on the LCD panel.

This sequence is repeated every 1000 ms, or every second.

The function `float tempC()` does the following:

Reads the output from the temperature sensor.

Uses two float variables: `tempVal` and `temp`.

`tempVal` takes the raw output from the pin A0 on the Arduino.

`temp` converts the `tempVal` to Celcius and returns it to the main program.

PART 2: DATA LOGGING ACTIVITIES

The next four activities in this chapter are parts of data logging. Once the data is logged, you can manipulate it in any way that is required.

The SD card shield mounts on top of the Arduino board. The shield has a sleeve for the SD card. We used a 2 GB card, which is plenty of memory for these activities. Remember, however, that not all SD card shields support SDHC cards. Figure 7.4 shows the SD card shield and the SD card, and Figure 7.5 shows the shield mounted on the Arduino, with the SD card inserted in its sleeve.

Figure 7.4
SD card shield with SD card.

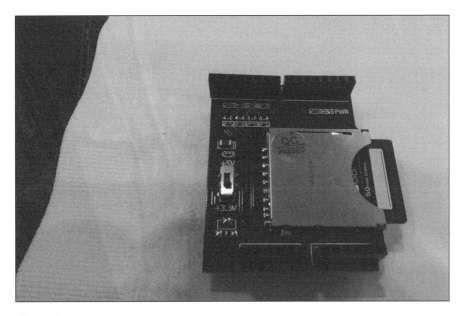

Figure 7.5
SD card shield mounted on Arduino.

The SD card library SD.h supports standard commands, which allow you to treat the card as you would treat a folder on your computer. You can create and delete files, write to files, and read from files.

Activity 1: Initializing the SD Card Reader

The SD card must be initialized before using the card. The sketch, along with explanatory notes, is given below. This example code is modified from the sketch available in the public domain at:

http://arduino.cc/en/Tutorial/CardInfo#.UxzL6fldW7o

The Hardware

Make sure that you have the SD card reader described in the list of components required for this chapter on hand. The SD card reader is a peripheral attached to the SPI (Serial Peripheral Interface) bus on the Arduino. Any peripheral device controlled by a microcontroller has four connections: MOSI (Master Out Slave In), MISO (Master In Slave Out),

CLK (Clock), and CS (Chip Select). These are connected to the Arduino pins as shown here:

MOSI	pin 11
MISO	pin 12
CLK	pin 13
CS	pin 10

You do not have to worry about these connections. The shield takes care of them for you.

Note

Note that even if it is not used as the CS pin, the hardware CS pin (10 on most Arduino boards, and 53 on the Mega) must be left as an output or the SD library functions will not work. These four pins on the Arduino (10–13) should not be used for any other purpose.

The Software: Writing the Sketch

Open the Arduino IDE and type the following sketch and save it as datalogger1_cardInit.

```
/*
  datalogger1_cardInit - Initializes SD card reader
*/
#include <SD.h>      // Library to use SD card reader
 const int chipSelect = 10;
void setup()
{
 // Open serial communications:
  Serial.begin(9600);

pinMode(10, OUTPUT); // make sure that the chip select pin is set to output

Serial.print("Initializing SD card...");
  // see if the card is present and can be initialized:
  if (!SD.begin(chipSelect))
  {
    Serial.println("Card failed, or not present");
    // don't do anything more:
    return;
  }
```

```
  Serial.println("card initialized.");
}
void loop()
{
  // Nothing to be done here
}
```

How the Code Works

This example code initializes the SD card reader.

The first line includes the library for using the SD card reader.

On the Arduino board, pin number 10 is used as Chip Select.

The `setup()` function does the following:

Opens the communication to exchange data between your computer and the Arduino board with a baud rate of 9600.

Pin number 10 is left open for output.

Print out "Initializing …"

The `if` condition checks if the SD card exists, and if it is does not, an error message is printed.

If the SD card does exist, "card initialized" is printed.

The `loop()` function in this sketch does nothing.

Activity 2: Writing Data to an SD Card

This activity describes how to write data to an SD card.

Setting Up Digital Data

We set up digital data on two of the Arduino pins, 5 and 6, as shown in Figure 7.6.

Figure 7.6
Data connections for digital data.

Connecting the Two Pins to:	Gives Values
GND on the Arduino	0, 0
5V	1, 1
One to GND and the other to 5V	1, 0 or 0, 1

Now type the following sketch, which writes the data to the SD card.

The Software: Writing the Sketch

Open the Arduino IDE and type the following sketch. Save it as dataLogger2_logData.

```
/*
dataLogger2_logData

This example shows how to log data from two digital pins
to an SD card using the SD library.

*/
```

```
#include <SD.h>
const int chipSelect = 10;
void setup()
{
  pinMode(5, OUTPUT);   // Pins used for data read by card
  pinMode(6, OUTPUT);
  pinMode(10, OUTPUT); // chipSelect pin as in the card_Init
  cardInit();        //Call the SD card Init function

  // Read the data into a string and store the string in a data file on the SD card
  // Make a string for assembling the data to log:
  String dataString = "";

  //Read the two digital data pins 5 and 6 and add the values to the dataString
  for (int dPin = 5; dPin < 7; dPin++)
    {
      dataString += digitalRead(dPin);
    }

  /* open the data file in the SD card. If the file is already there,
    it is opened, and what you store will be added to the end of the file.
    If the file is not there, it is created. The FILE_WRITE parameter opens the
file for writing.
    If this parameter is not given, file is open for Read only.
    Note that only one file can be open at a time, so you have to close this one before opening
another.
    */

  File dataFile = SD.open("01Mar12.txt", FILE_WRITE);

  // if the file is available, write to it:
  if (dataFile) {
    dataFile.println(dataString);
    dataFile.close();
    // print to the serial port too:
    Serial.println(dataString);
  }
  // if the file isn't open, pop up an error:
  else {
    Serial.println("error opening 01Mar12.txt");
  }
}

void loop()
{
  // Nothing to be done in an endless loop
}
```

```
void cardInit()
{
  // Open serial communications:
  Serial.begin(9600);
  Serial.print("Initializing SD card...");

  // see if the card is present and can be initialized:
  if (!SD.begin(chipSelect))
  {
    Serial.println("Card failed, or not present");
    // don't do anything more:
    return;
  }
  Serial.println("card initialized.");

}
```

How the Code Works

This example illustrates logging the data to an SD card using the values at two digital output pins on the Arduino. This example code is modified from the sketch available in the public domain at:

http://arduino.cc/en/Tutorial/CardInfo#.UxzL6fldW7o

Revisit the sketch datalogger1_cardInit for details on setting up the SD card shield. All these steps have been included in a function card_Init, which is called in the setup() function.

Note

Before you run the sketch, connect the Arduino pins 5 and 6 to either GND or 5V, in any combination you like—for example, pin 5 to GND and pin 6 to 5V. Or, both pins can be connected to GN. This will be the data you will write to the file on the SD card.

The SD.h file is imported and pin 10 is again left open as in Activity 1.

The setup() function sets pins 5 and 6 to 0 or 1 at the beginning of the activity.

cardInit() calls the SD card Initialize function.

A string variable is created to hold the values from the pins.

A for loop is used to read the values from pins 5 and 6 and build the dataString.

If the `datafile` exists, the values from the `dataString` are written to the `datafile`, and the file is closed. The contents are also printed via the `serial.println` command.

The `loop()` function is a blank function; in this program it does nothing.

The `cardInit()` function does the following:

Opens the serial port for communication and initializes the SD card.

If the SD card is not found, an error message is printed.

If the SD card is found, "card initialized" is printed.

Activity 3: Reading Data from a File

In this activity, you will read the data from the file in which you stored the data in Activity 2.

Writing the Sketch

Open the Arduino IDE and type the following sketch. Save it as dataLogger3_readData.

```
/*
 dataLogger3_readData - Read data from a file
 This example shows how to read data from an SD card using the SD library
 */
#include <SD.h>
const int chipSelect = 10;
void setup()
{
  pinMode(10, OUTPUT);
  cardInit();   //Initialize the card
  // Open the file. Note that only one file can be open at a time,
  // so you have to close this one before opening another.
  File dataFile = SD.open("01Mar12.txt");
  // This filename should be identical to the one you created in Activity 2.
  if (dataFile)
  {
    Serial.println("01Mar12.txt:");
    // read from the file until there's nothing else in it:
    while (dataFile.available())
    {
        Serial.write(dataFile.read());
    }
    // close the file:
    dataFile.close();
  }
```

```
  else
  {
    // if the file didn't open, print an error:
    Serial.println("error opening 01Mar12.txt");
  }
}
void loop()
{
  //
}

void cardInit()
{
  // Open serial communications:
  Serial.begin(9600);
  Serial.print("Initializing SD card...");
  // see if the card is present and can be initialized:
  if (!SD.begin(chipSelect))
  {
    Serial.println("Card failed, or not present");
    // don't do anything more:
    return;
  }
  Serial.println("card initialized.");
}
```

How the Code Works

The SD.h library is included.

The setup() function does the following:

Initializes the SD card.

Opens the file 01Mar12.txt and assigns it to the dataFile variable.

If the file exists, then the serial.println command prints the name of the file: 01Mar12.txt

The while loop checks if the file exists, and if so, it reads the contents of the file and writes the content to the serial monitor.

The dataFile is closed

Otherwise, an error message is printed to the serial monitor.

The loop() function is empty and it does nothing in this example.

The cardInit() function does the following:

Opens the serial port for communication and initializes the SD card.

If the SD card is not found, an error message is printed.

If the SD card is found, "card initialized" is printed.

Activity 4: Logging Temperature Data

In this activity, all the data logging parts come together. Temperature data is taken over a period of time in regular intervals, and is logged to a file on the SD card.

Hardware Connections

You will have to repeat the connections to the TMP36 and the LCD panel, with the SD card shield present. First, remove the two wires that you connected to pins 5 and 6 on the Arduino for the previous activity. You will notice that even with the shield mounted the Arduino pin numbers have not changed, and can be seen on the shield itself. So you can make the same connections for the TMP36 and the LCD panel. The total layout is seen in Figure 7.7 in all its glory.

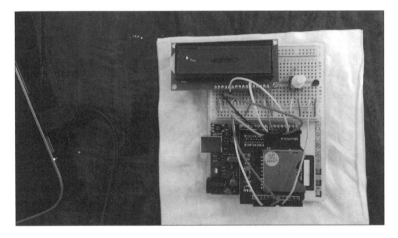

Figure 7.7
Completed layout for temperature data logging.

Writing the Sketch

The following sketch accomplishes the task of reading the temperature from the TMP36, displaying the temperature on the LCD panel, and saving the temperature data to a file on the SD card. Open the Arduino IDE and type the following sketch.

```
/* Temperature sensor data displayed on LCD panel, logged to SD card
*/
// include the library code for the LCD display:
#include <LiquidCrystal.h>
#include <SD.h>       //Library for SD card
// initialize the library with the numbers of the interface pins
LiquidCrystal lcd(7, 8, 9, 4, 5, 6);
const int chipSelect = 10;
float tCel;
void setup()
{
  Serial.begin(9600);
  // set up the LCD's number of columns and rows:
  lcd.begin(16, 2);
  // Print a message to the LCD.
  lcd.print("Temperature: ");
  pinMode(10, OUTPUT);
  cardInit();      //Call the SD card Init function
  File dataFile = SD.open("01Mar12.txt", FILE_WRITE);   //Open a data file

  for (int j = 0; j < 20; j++)
  {
    tCel = tempC(); // Read the temperature

    // LCD Display: set the cursor to column 0, line 1
    // (note that line 1 is the second row, since counting begins with 0):
    lcd.setCursor(0, 1);
    lcd.print(tCel);
    lcd.println(" C");

    //Send the datastring to the file
    if (dataFile)
    {
      dataFile.print(tCel);
      dataFile.println(" C");
      Serial.print(tCel);
      Serial.println(" C");
    }
    else
    {
      Serial.println("error opening 01Mar12.txt ");
    }
    delay(1000);
  }
```

```
  dataFile.close();
}
void loop()
{
}
void cardInit()
{
  // Open serial communications:
  Serial.begin(9600);
  Serial.print("Initializing SD card...");
  // see if the card is present and can be initialized:
  if (!SD.begin(chipSelect))
  {
    Serial.println("Card failed, or not present");
    // don't do anything more:
    return;
  }
  Serial.println("card initialized.");
}
float tempC()
{
  //Read the temperature
  float tempVal, temp;
  tempVal = analogRead(A0); // This is the raw output from the TMP36 sensor
  temp = ((tempVal*5000/1024)-500)/10; // Raw data converted to Celsius degrees
  return temp;
}
```

How the Code Works

Include the libraries for the SD card reader and the LCD panel.

The LCD panel is initialized with the interface pin numbers 7, 8, 9, 4, 5, and 6.

Variables are used.

The constant int chipSelect is used by the SD card reader.

A float tCel is used to hold the temperature value.

The setup() function does the following:

Starts the serial monitor with a baud rate of 9600.

Opens the LCD panel with 16 rows and 2 columns.

Prints the title "Temperature:" on the LCD panel.

Sets pin 10 for output.

The `for` loop does the following for up to 20 values (this can be changed per your requirements):

Calls the function `tempC()` and assigns the value to `tCel`.

Sets the cursor to column 0, row 1.

Prints the value of `tCel`.

Prints the string " C".

Checks if the `dataFile` exists, and if so the contents from `tCel` are written to the `dataFile`, else an error message is written on the serial monitor.

A delay of 1000 milliseconds is set so the values can be recorded at intervals of one second.

The `loop()` function again does nothing as in the earlier examples.

The function `tempC()` reads the temperature from the temperature sensor. Two float variables `tempVal` and `temp` are used.

`tempVal` takes the raw output from the pin A0 on the Arduino.

The `temp` variable takes the converted Celsius value from the `tempVal` and returns it to the main program.

CONCLUSION

In this chapter, you learned how to use an SD card reader to log data. You also learned how to use a temperature sensor and an LCD panel to display messages and data.

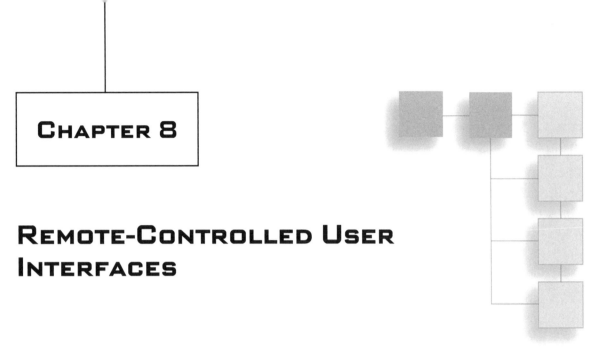

CHAPTER 8

REMOTE-CONTROLLED USER INTERFACES

Graphical user interfaces (GUIs) provide the primary interaction between general consumers and the computers we use. In this chapter, we discuss the different platforms available to create user interfaces and implement an interface to control your Arduino robot over Wi-Fi.

CHAPTER OBJECTIVES

- Install and learn the NetBeans IDE
- Learn basics of the Java programming language
- Connect to your Arduino via Java
- Create a graphical Java application using JavaFX
- Control your robot remotely

INTRODUCTION

So far, you have learned the ins and outs of Arduino development, including design, fabrication, and programming. By now, you've noticed that at the low levels of robot operation, you interacted with data in a text-based environment; there are no pictures or videos! You programmed by typing text, you sent commands via text, and you even had to use a terminal or text console on your computer to communicate with your wireless robot! This is how all computers operated until the 1980s, when Xerox and SGI pioneered the concept of a "desktop," complete with a mouse to click on buttons. At the time, these machines were not available to the public and sold for more than $25,000. The graphical user interface (GUI) was then revolutionized by Apple's Lisa computer and primarily by Macintosh, the first commercially successful computer to use a true GUI.

These interfaces, however, were nothing compared to today's standards. In an age with multi-touch smartphones, complex Internet browsers, and incredibly powerful processing units, untold functions are now possible, and the concepts of "design" and "user experience" are at the forefront of software development.

Development Software

With numerous software platforms available to develop on, one of the biggest decisions for developers is which language, and which relevant libraries, they can use to develop their application, to optimize speed, functionality, development time, and "look and feel." While most software projects utilize a designer whose whole purpose is to design how the user will experience the application, much of the implementation of these designs falls on the software developer. Following is an exploration of a few of the front-runners of software languages that have remarkable GUI tools.

Java

Java, one of the most common programming languages, provides a structured and very readable syntax that allows quick development and easy debugging. As an object-oriented language with millions of libraries available, it can provide any function required. Java is medium-slow by software standards, but it is so powerful that it is used for systems all over the world, from servers to microwave controllers. Java contains two primary UI libraries, called Swing (previously Abstract Window Toolkit) and the more recent library, JavaFX. These libraries provide a layer between the functional code and the view of the interface, allowing the UI to be designed with no code required. The libraries then attach the code to the designed components. These tools are most suited for desktop applications but can also be put online through Java applets (see Figure 8.1). Android phone applications are built using a form of Java.

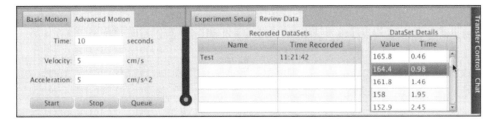

Figure 8.1
Java applet example.
Source: Oracle JDK.

JavaScript/Web User Interfaces

Not to be confused with Java, JavaScript (JS) is a web technology developed by Netscape. It was named JavaScript to gain from the hype created by the Java language. An example of JavaScript is shown in Figure 8.2. While JavaScript cannot be used alone, it is responsible for the majority of modern interactive websites. JavaScript is used to modify HTML documents in real time, which when styled with CSS produce some of the best-looking applications and web interfaces available. Unlike Java and traditional programming languages, JS is not compiled; it is interpreted by the browser. This allows JS to be updated easily, but this can lead to security issues. It is also necessary to use other technologies, such as AJAX and PHP, to connect to external resources.

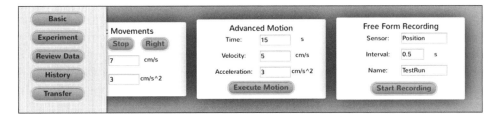

Figure 8.2
JavaScript example.
Source: RILE Inc.

Python

Python is another very high-level language, which is used mostly by the scientific community for its simplicity and excellent mathematics libraries. Python also has a complete UI package that allows very simple UI management. Python is desktop based, and is a cross between compiled languages and interpreted languages, as both features can be used. Python is slower than Java and not as widely used; however, its use is expanding rapidly.

Objective C

Objective C is a language based on C++ and designed for the Apple iOS family of devices. Apple has packaged an incredible suite of styled UI components that can be used right out of the box. As a completely compiled language, Objective C is incredibly fast, even on mobile devices, while powerful enough to implement 3D gaming engines. While arguably the most difficult of these languages, it is also specialized for iOS devices. However, this has not stopped Objective C programmers from having the highest income, on average, of any software developers.

While there are numerous other UI platforms available, such as QT, MatLab, and more, these provide a sample of the most prevalent technologies. Each language has its benefits and disadvantages, and sometimes it is necessary to mix and match components to create the most efficient and well designed system. In most cases, there are at least two or three systems that work together to deliver the application to the user, including a UI, a server, and a database. These components are mapped out prior to the start of programming.

A Graphical Control for Arduino

In this project, you will create a graphical user interface using the Java platform, which will connect via the Wi-Fi connection you created in Chapter 5. You will then add functions into the robot control loop (which we will copy from Chapter 5) to move the robot forward or backward and turn on your command. As with any multi-tier software project, we will begin with a diagram (see Figure 8.3).

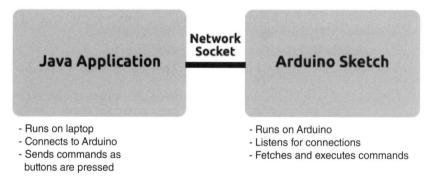

Figure 8.3
System diagram for interfacing Arduino with Java.

As you can see, we have two primary chunks of code: one running the UI, and one running the Arduino. The two are connected over a network socket that creates a link between the two devices. Data will then be sent between the two devices. We will then create a UI using JavaFX to connect to the Arduino and send/receive commands.

MATERIALS REQUIRED

- Standard Arduino robot used in previous chapters
- Arduino Wi-Fi shield
- USB mini cable
- Ultrasonic range finder
- Wi-Fi network
- Computer with Internet access
- NetBeans with JDK (free software)

INTRODUCTION TO JAVA

So far, within this book, you have written all your sketches in the Arduino language—a custom platform for the Arduino board. While Java is very similar to Arduino, Java is an object-oriented language, which is very different. However, because we will use limited Java features, it will seem familiar. So let's jump in. The following is a basic Java program.

```java
import java.util.ArrayList;
/**
 * Created by Alex Whiteside on 5/18/14.
 */
public class basicClass {

    public static void main(String[] args) {

        System.out.println("Hello World!");

    }
}
```

First, let's look at the similarities:

1. Comments are shown using / and * notation. // is still a valid single line comment.

2. Functions (methods in Java) are encased with {} and arguments are given using ().

3. Dot notation, or using periods to call parts or functions are still used.

4. " " are used to denote strings of text.

5. Semicolons are still used to end each command.

Now, look at the differences:

1. In Java, everything needs to be nested in a class—in this case, basicClass.

2. In Java, there are no specified setup() and loop() functions, but there is a main function that signals the start of the program.

3. Instead of #include, Java uses import statements to add libraries.

In order to see what this program does, we'll need to install some special software as described in the following section.

Downloading and Installing NetBeans

To compile and run Java, you will need the Java Development Kit (JDK), which compiles .java files into Java byte code that can be run on your system. We will install an IDE (Integrated Development Environment) called NetBeans, which includes this JDK as well as an environment that makes it easy to compile and run Java code, much like the Arduino IDE.

1. To download NetBeans, either grab the installer from our companion website (www .cengageptr.com/downloads) or navigate to https://netbeans.org/downloads/index .html, and select the Java SE version. Click the Download button to initiate the download.

2. When the download is complete, double-click the installer and follow the onscreen instructions to install NetBeans.

3. Once installed, you will see a screen similar to that shown in Figure 8.4. The top left panel contains your active projects, the main window will contain your code, and the bottom of the screen displays the output of your programs.

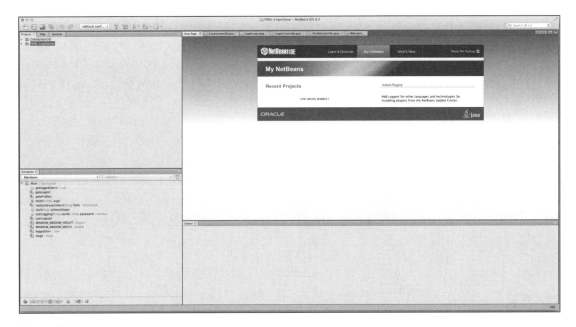

Figure 8.4
NetBeans window.
Source: Oracle NetBeans.

4. Create a new project by selecting File > New Project. Select the Java category and double-click on Java Application.

5. Name this application Network Demo and press the Finish button.
   A program outline containing most of the demo program has been created for you. Within the `main` method, add this line of code:

   ```
   System.out.println("Hello World!");
   ```

6. Select the Run button (the green arrow) on the top toolbar, or from the File menu, select Run > Run Project.

7. In the console window at the bottom, you will see the words "Hello World!" printed to the screen. If there is an error and these do not print, check your installation settings for NetBeans; you may need to re-install.

NETWORKING IN JAVA

Our program's principle function is to connect to the Arduino over Wi-Fi and send/receive commands. We will do this over a network socket. Network sockets are language-agnostic and can be used to interface systems in real time over a physical connection. Much of this functionality is done by the operating system, which sends your data over physical link layers to the other device, which then translates the data for that system. The two important things needed to create a network socket are a connection IP address and port. Each time you connect to a network, an IP address is established for other devices to communicate with your device. Ports are internal to your device and allow multiple connections to happen to/from your device. They are simply numerical identifiers and can be (almost) whatever you want, provided another service is not already using that port. For example, web servers default to port 80, while MySQL databases are on port 3306. We have been using the Telnet port 23 for communication with the Arduino, and we'll continue to use it. You create a socket using the following line:

```
Socket soc = new Socket("IP Address", port);
```

Where `Socket` defines that this is a socket, `soc` is the variable holding the socket connection, `IP Address` is the IP address of the Arduino, and `port` is the port to which you will connect.

The second library that will be used is that of InputStreams and OutputStreams, which allow you to send plain text across the socket. You can get handles to these streams from the socket object, using dot notation:

```
InputStream in = soc.getInputStream();
OutputStream out = soc.getOutputStream();
```

You can then read and write from these streams:

```
char input = (char) in.read();
out.write("Hello".getBytes());
```

Let's try these out.

1. Open your code from Chapter 5, "Robot Networking and Communications with Wi-Fi," and compile it onto the Arduino. Open the serial monitor and make sure it connects to your network. Note the IP address.

2. Within your Network Demo program, in the `main` method, create the socket as follows:

   ```
   Socket soc = new Socket("10.0.1.53",23);
   ```

Replace 10.0.1.53 with your Arduino's IP address from the serial monitor.

3. Red lines should appear under the word Socket, which means you have not imported the Socket library. To do so, right-click on the word Socket and select Fix Imports. This will automatically add the line `import java.net.Socket;` to your program. You can also click on the red light bulb on the left of the screen to show possible fixes.

4. There is still a red line under the rest of the line, indicating an exception. Because there are numerous things that could go wrong when communicating over a network, Java requires you to handle such things. For now, we will request that Java print said errors and terminate the program by modifying the `main` method header to read:

```
public static void main(String[] args) throws Exception {
```

5. We will now get the relevant streams by adding the following lines:

```
InputStream in = soc.getInputStream();
OutputStream out = soc.getOutputStream();
```

6. For testing, we will need some way of sending data to the console, so we will therefore create a Scanner that will allow us to import text:

```
Scanner keyboard = new Scanner(System.in);
```

7. The next step is to print all incoming data and write all outgoing data. We want to do this continually, so we will create a while loop that remains true:

```
while(true){


}
```

8. Inside the while loop, we will print all incoming characters:

```
System.out.print((char)in.read());
```

This line calls the read function from the input stream, returning a numerical representation of a character if it is available, or nothing if none is available. The `(char)` code converts this `int` to a `char` unit, which can be printed by `System.out.print();`

9. We also want to write data if we send it, but we may want to send more than one character at a time, so we will use the concept of lines to determine what to send. A line is created when you strike the Enter key using the following conditional:

```
if(keyboard.hasNextLine())
{

}
```

10. Within the `if` statement, write the data using the following line:

```
out.write(keyboard.nextLine().getBytes());
```

In Java, nested functions happen first, causing the scanner to retrieve the line entered and convert it to a byte array that is written to the output stream.

11. In total, your program should look like this:

```
import java.io.InputStream;
import java.io.OutputStream;
import java.net.Socket;
import java.util.Scanner;

/**
 *
 * @author Alex Whiteside
 */
public class NetworkDemo {

    /**
     * @param args the command line arguments
     */
    public static void main(String[] args) throws Exception {
        System.out.println("Hello World!");
        Socket soc = new Socket("172.16.42.7",23);
        InputStream in = soc.getInputStream();
        OutputStream out = soc.getOutputStream();
        Scanner keyboard = new Scanner(System.in);
        while(true)
        {
            while(in.available()==0)
            System.out.print((char)in.read());
            if(keyboard.hasNextLine())
            {
                out.write(keyboard.nextLine().getBytes());
            }


        }


    }

}
```

12. Run this application by selecting the green Run arrow in the toolbar. Ensure that your Arduino is running the correct software. You will now be able to enter characters to send to the Arduino and receive the data sent from the board. This is exactly like the Telnet program you ran in the previous chapter; however, you now have control over this code.

CREATING THE USER INTERFACE

Now that you have an underlying platform to communicate with the Arduino, we will construct a user interface to allow this interaction to be like a regular game controller.

1. Let's rename our NetworkDemo class to something more appropriate: "ArduinoCommunicator." Right-click on the NetworkDemo Class in the projects tab on the left, hover over Refactor, and then select Rename. Enter "ArduinoCommunicator" and press OK.

2. Create a JFrame Form by right-clicking on the package "network.demo" in the projects pane and selecting New > JFrame Form. Name this class "ArduinoForm." A JFrame Editor should appear, as shown in Figure 8.5.

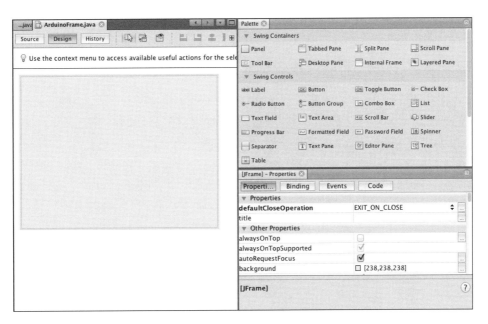

Figure 8.5
Java JFrame Editor.
Source: Oracle NetBeans.

Let's locate some important components in the JFrame Editor. In the top right is a Palette tab, which contains all the physical components that can be added to your interface, including buttons, labels, and text areas. Underneath the Palette tab is the Properties tab, which contains the properties of whatever you select. On the left side is the designer, which displays a gray box, which is the frame of your interface. This is where you will drag your components to create them. Notice the Source and Design tabs at the top of this canvas. You will toggle between these two tabs to edit the design of the interface and to modify the source code to create actions for your buttons.

3. Create a button by dragging a button from the Palette to the canvas. "jButton1" will appear on the canvas. Select the button in the canvas and locate the Text property in the Properties tab. Double-click on the "jButton1" text and change it to "Forward." The button's name should change in the canvas. In the Properties tab, select Code and change the Variable Name to "forwardButton."

   Notice that you can drag the button to wherever you like, and as you add more components, they will add themselves relative to this component.

4. To describe what this button will do, double-click the button. A function (method) will be created in the frame class called forwardButtonActionPerformed. The code inside this function will be called whenever the user clicks the button in the UI. Enter the following line:

   ```
   System.out.println("You Pressed A Button");
   ```

5. Let's run this frame. In the Project tab, select ArduinoFrame.java, right-click and then select Run File. You should see a frame appear, as in Figure 8.6.

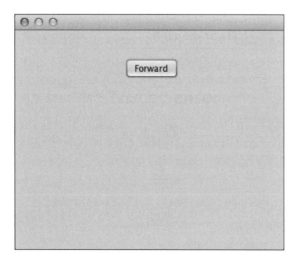

Figure 8.6
An example JFrame.
Source: Oracle NetBeans.

6. Click the button in your interface. Notice that in the Output tab at the bottom, the system prints your message. We will use this concept to tell the ArduinoCommunicator to tell the robot to move whenever you click a button.

Now that we understand all three parts of the system (robot, controller, and interface), let's develop all the functionality on the robot and build up the system. Fire up your Arduino IDE and load your Wi-Fi project onto the board.

1. Let's start by adding the necessary requirements to control the motors to make the robot run. As in previous chapters, add the following lines before the setup() function:

```
#define CW 0
#define CCW 1
#define MOTOR_A 0
#define MOTOR_B 1
const byte PWMA = 3;    //  PWM control (speed) for motor A
const byte PWMB = 11;   //  PWM control (speed) for motor B
const byte DIRA = 12;   //  Direction control for motor A
const byte DIRB = 13;   //  Direction control for motor B
```

2. We will now set the pins appropriately by adding the following lines to the start of the setup() function:

```
//  All pins should be set up as outputs:
pinMode(PWMA, OUTPUT);
pinMode(PWMB, OUTPUT);
pinMode(DIRA, OUTPUT);
pinMode(DIRB, OUTPUT);

//  Initialize all pins as low:
digitalWrite(PWMA, LOW);
digitalWrite(PWMB, LOW);
digitalWrite(DIRA, LOW);
digitalWrite(DIRB, LOW);
```

3. Let's create five functions to provide complete motion for the robot: forward, backward, left, right, and stop. These work similarly to those you have written throughout this book.

```
void forward()
{
digitalWrite(DIRA,CW);
digitalWrite(DIRB,CW);
```

```
analogWrite(PWMA,150);
analogWrite(PWMB,150);

}


void turnRight()
{

digitalWrite(DIRA,CCW);
digitalWrite(DIRB,CW);
analogWrite(PWMA,150);
analogWrite(PWMB,150);
}
```

4. Complete the remaining three functions and test them by calling them in the setup() function to ensure they work in your robot setup.

We will now assign each of these actions to an op-code, or identifier, which will allow us to call them from the remote computer easily, just as we did for the sonars with the 'u' command. We shall designate the commands as listed in Table 8.1.

Table 8.1   Op-Code Assignment

Op-Code	Function
'f'	forward()
'b'	backward()
's'	stop()
'l'	left()
'r'	right()

1. Locate the loop() function where you receive data from the client.

2. Using 'u' as a template, create commands for each of the functions in a series of if statements. For example:

```
if(temp =='f')
    {

    forward();

    }
```

Create these commands for each of the functions.

We will now create relevant functions in the ArduinoCommunicator, which will be called by the interface. We will move our instructions out of the `main` method and into methods that can be called by the buttons' event handlers.

1. Return to NetBeans and open the ArdinoCommunicator class.

2. We will define global variables for the socket input and output streams so that they can be used throughout the class. Below the line that reads:

```
public class ArduinoCommunicator
```

add the following lines:

```
Socket soc;
InputStream in;
OutputStream out;
```

3. We will create a constructor that allows us to create the ArduinoCommunicator similarly to how the `setup()` function allows you to set up the Arduino. The constructor must be named the same as the class and goes just below the variables you just defined.

```
public ArduinoCommunicator()
    {  }
```

4. Inside this constructor, we will set up the socket and streams as follows:

```
public ArduinoCommunicator() throws Exception
  {

      soc = new Socket("192.168.1.60",23);
      in = soc.getInputStream();
      out = soc.getOutputStream();


  }
```

5. We can now remove the `main` method entirely from the class.

6. We will now create methods to send each of the op-codes from the user interface. The methods should look like this:

```
public void moveForward() throws Exception
    {

        out.write('f');


    }
```

Create methods for each of the op-codes, naming them `moveForward()`, `moveBackward()`, `turnLeft()`, `turnRight()`, and `stop()`. In each method, write the op-code character to the Arduino.

Note

In each of these cases, we must add `throws Exception` to the header. Exceptions are Java's form of errors. In most cases, you should handle the error by reconnecting.

We will now complete the actual interface and connect it to the Arduino.

1. Open the ArduinoFrame class and select the Source button at the top.
   Locate the class header at the top of the file, where you placed the variables earlier. Create a variable in the ArduinoFrame for an ArduinoCommunicator by adding the following line:

   ```
   ArduinoCommunicator arduino;
   ```

2. In the ArduinoFrame constructor, add the following line before the call to the `initComponents()` method:

   ```
   arduino = new ArduinoCommunicator();
   ```

3. The start of your class should look like the following:

   ```
   public class ArduinoFrame extends javax.swing.JFrame {

       ArduinoCommunicator arduino;

       /**
        * Creates new form networkFrame
        */
       public ArduinoFrame() {

   try{
           arduino = new ArduinoCommunicator();
           initComponents();
       }catch(Exception e){e.printStackTrace();}
   }
   ```

4. Click the Design button at the top of the editor. The Palette should become visible. Drag four more buttons to the canvas so it looks like Figure 8.7.

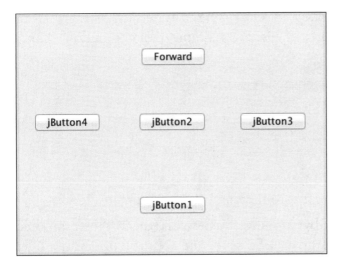

Figure 8.7
Populated JFrame.
Source: Oracle NetBeans.

5. Rename the buttons so that the names simulate a gamepad. The button names should be changed to resemble those in Figure 8.8.

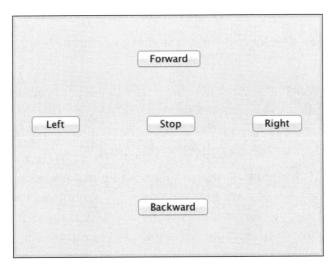

Figure 8.8
JFrame interface with new button names.
Source: Oracle NetBeans.

6. Double-click each button to create an event listener. Inside each listener, call the correct method from the arduino variable. For example, the Stop button's handler will look like this:

```
private void stopActionPerformed(java.awt.event.ActionEvent evt) {
    try{
        arduino.stop();
    }catch(Exception e){}
}
```

Create these for each button. Notice the try{}catch(Exception e){e.printStackTrace();} This is required, and will print an error if something incorrect happens.

We will now test the functionality by running the systems. Since the communication occurs via Wi-Fi, you can still connect over USB serial and see the output of the Arduino.

1. Start the Arduino by powering it on and opening the serial monitor.

2. Wait until the Arduino has reported a connection and an IP address, and then run the ArduinoFrame class by right-clicking on it and selecting Run File. Your interface will appear, and can be used to control the robot.

OPTIONAL ACTIVITIES

Following are a few ideas for further developing what you have accomplished in this chapter:

■ **Improve the Graphic Design:** We've just scratched the surface of GUI development and design. The NetBeans IDE allows complex interaction and design of Palette objects into beautiful interfaces that are fun and easy to use. Try experimenting with different icons, colors, and designs to create a better-looking GUI!

■ **Add Other Arduino Sensors:** At this point, you are only sending movement commands over the network; however, you are not limited to this. In future chapters, we explore bi-directional communication over the network. Try using this skill to retrieve sensor data over the network and display it in your user interface.

■ **Print Interface State:** Unless you have a terminal in front of you like in the Arduino IDE, you don't know if anything has gone wrong. Add a JLabel or other indicator at the top of the GUI to print the connection status of the Arduino.

CONCLUSION

You have now seen a basic way to create an interface to the Arduino outside of the autonomous algorithms you have created. This involves multiple languages, devices, and connections, but simulates the real world in a much better way. The world we live in is composed of numerous systems, devices, and people who must communicate to get things done!

The complete code for both the Arduino and Network Interface can be found on the companion website (www.cengageptr.com/downloads).

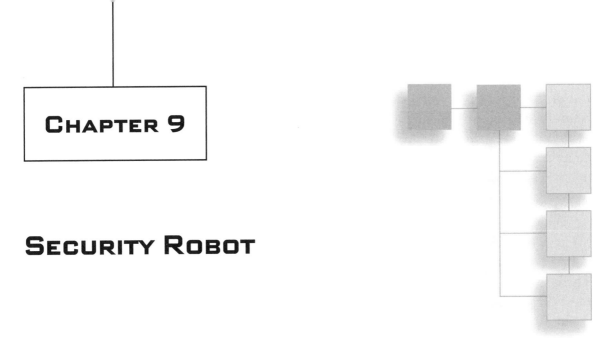

CHAPTER 9

SECURITY ROBOT

Cell phones, pocket-sized cameras, and web cams have had a huge impact on our daily lives. In fact, researchers estimate that as of 2011, we take more pictures in 2 minutes than all photos taken in the entire 19th century, totaling close to 400 billion photos per year, and rising. Cameras personalize technology and provide a primary means of accessing remote spaces, whether over Skype, Google Maps, or e-mail. This chapter explores digital camera technology and provides you with the means to create a mobile camera robot, which can be used for remote security.

CHAPTER OBJECTIVES

- Install and configure the RadioShack Arduino camera
- Modify RadioShack software to record all images
- Review images on a computer

INTRODUCTION

Home security systems are becoming increasingly popular as many companies are using technology to provide better security for drastically lower prices, simply by automating tasks typically done by customer service representatives. Your Arduino robot can join this revolution through the addition of a camera and a few fancy programming concepts. In this chapter, we will take advantage of RadioShack's Arduino camera and modify their software to create a motion-detecting security camera.

We'll focus on the camera hardware and how the camera communicates with the Arduino. Digital cameras are composed of three critical components: the lens, the sensor, and the processor. The lens of a camera is responsible for focusing the light into a narrow beam for the sensor to read. Lenses come in a variety of sizes, shapes, and costs; some are as cheap as a few cents, small, and produce blurry images, while others weigh 2 pounds and contain multiple elements, resulting in perfect shots. Lenses are defined by their focal length (zoom) and maximum aperture (how wide the lens can open).

The sensor of the digital camera is essentially an array of components that get excited by light focused from the lens. This excitement creates electrical charges that are fed into the processor. While there are numerous types of sensors, the most common are CCD (Charge-Coupled Device) and CMOS (Complementary Metal-Oxide Semiconductor). These sensors vary in size and typically provide the "mega-pixel" specification you see on most digital cameras. For typical digital cameras, they also provide the maximum shutter speed. In DSLRs (Digital Single-Lens Reflex—a more professional type of camera), a mirror assembly physically moves out of the way of the sensor, producing a time-specific exposure.

The processor of the camera takes the information generated by the sensor (typically an array of light values) and uses a profile for that lens to convert the image values into an accurate picture. This image is then sent over to a microprocessor to be saved, streamed, or analyzed.

In this project, we will use RadioShack's camera hardware to set up a motion-detection program for their camera and record data from the camera every time motion is detected. We will save the data (images) to the SD card, which can be further viewed on the computer.

MATERIALS REQUIRED

- Standard Arduino robot used in previous chapters
- RadioShack camera shield
- SD card shield with SD card
- Computer with SD card reader

SETTING UP THE HARDWARE

This project relies on multiple shields working together to capture images and send them to the computer: the camera shield to take the photo, and the SD shield to save the data.

1. Connect the SD card shield to the top of the Arduino. Make sure an SD card is inserted into the SD card shield.

2. Download the RadioShack camera shield files from: http://blog.radioshack.com/uploads/2013/01/RadioShack-Camera-Board-Support-Files.zip

3. Following the diagrams in the documentation, connect the camera to the SD card shield using the UART configuration with the included cable. Check your connections for the shields by comparing them with Figure 9.1.

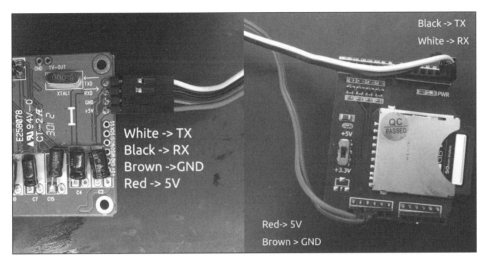

Figure 9.1
Shield connections.

We'll now test the camera shield using the included sketch to ensure that everything is connected properly. When programming, it may be necessary to remove the shield/camera assembly, since they use the serial ports used by the programmer.

1. Open the camera.ino sketch inside the included camera files and load it into the Arduino IDE.

2. You will need to install the SDF and SD card libraries to interface with the camera shield's sample program. You can download those from https://sdfatlib.googlecode .com/files/sdfatlib20131225.zip, or from the companion website (www.cengageptr .com/downloads), or from the RadioShack support files. Installing libraries can be done by selecting Import Library and adding the download to the IDE. The include calls are already done for you.

3. Compile and download this code into your Arduino Uno, replace the shield assembly, and reset your Arduino. The installed software is a motion detector, so simply waving your hand in front of the camera will activate it. Do this a few times to ensure that an image has been captured.

4. Insert the SD card into your computer or card reader and open the main directory. You should see an image called temp.jpg. Open this image to see a (probably blurry) image of yourself (Figure 9.2). Don't worry; we'll correct the image quality later. If you do not see any images, ensure that the program compiled correctly and repeat the process.

Figure 9.2
The blurry "temp.jpg" image.

We'll now modify this program to allow us to save multiple images instead of just over-writing temp.jpeg. In a typical computer, we would use the current time to timestamp the image, so there would be more information about it. Unfortunately, Arduino does not have a large enough timer to keep an accurate date and time, so we must use a generic counter. We'll increment it for each image that's saved, so you will end up with a series of images: image1.jpg, image2.jpg, image3.jpg, and so on.

1. First, open up the camera sketch provided by RadioShack.

2. Create a global variable to act as our counter. We'll call it imageCount and initialize it to 0 just before the setup() function with the following line:

```
int imageCount = 0;
```

3. We'll also create two string variables to hold the base name and the file extension.

```
String imageName = "image";
String imageEnding = ".jpg";
```

4. We must then modify the `capture_photo()` function to save the image as a dynamic file name. Find the `void capture_photo()` function near the bottom of the file. Immediately after the start of the function, create a string variable to hold the file name:

```
String fileName;
```

Create the `fileName` by combining the `imageName`, `imageCount` and `imageEndings`.

```
String fileName = imageName + imageCount + imageEnding;
```

5. We will also have to convert the string into a `char` array in order for it to be used by all Arduino functions.

```
char fileNameChar[50];
fileName.toCharArray(fileNameChar,50);
```

6. We'll now increment the counter so the next time we record a photo, it uses the next number.

```
imageCount = imageCount +1;
```

7. To feed our `fileNameChar` into the recording code, we'll replace every instance of `temp.jpg` with `fileNameChar`. First, change the `if` statement to remove the file, if it exists, to:

```
if(sd.exists(fileNameChar)) sd.remove(fileNameChar);
```

8. On the next line, replace `temp.jpg` with `fileNameChar` during the call to the `open()` function.

```
if (!myFile.open(fileNameChar, O_RDWR | O_CREAT | O_AT_END)) {
```

9. The last change is the next call to `myFile.open()` halfway through the function. Replace it with:

```
myFile.open(myFileChar, O_RDWR);
```

Now it's time to test. Download your software to your Arduino. (You may need to remove the shield.) Plug the device into a power source (or connect via USB) and wave your hand over the lens of the camera. Wave a few more times each minute to create motion.

After a few minutes, turn off the Arduino and remove the SD card. Insert it into your computer or card reader and open the folder. You will see a number of images in numerical order. Double-click one to open it and ensure that you can see the picture (Figure 9.3).

Name	▲	Size	Kind
TEMP.JPG		Zero bytes	JPEG image
TEMP0.JPG		13 KB	JPEG image
TEMP4.JPG		12 KB	JPEG image
TEMP5.JPG		13 KB	JPEG image
TEMP7.JPG		12 KB	JPEG image
TEMP9.JPG		12 KB	JPEG image
TEMP13.JPG		13 KB	JPEG image
TEMP22.JPG		Zero bytes	JPEG image
TEMP35.JPG		12 KB	JPEG image
TEMP61.JPG		12 KB	JPEG image
TEMP63.JPG		12 KB	JPEG image
TEMP65.JPG		12 KB	JPEG image
TEMP67.JPG		12 KB	JPEG image
TEMP69.JPG		12 KB	JPEG image
TEMP71.JPG		Zero bytes	JPEG image

Figure 9.3
Multiple images stored on the SD card.

You'll notice that all the images are blurry. This is due to the lens' focus. To correct this, twist the lens on the camera. For close objects, twist the lens counter-clockwise to loosen it and move it farther from the sensor. For farther objects, twist the lens clockwise to tighten it and move it closer to the sensor.

THE COMPLETE RELEVANT CODE BLOCKS

For reference, following are the complete code blocks for the variable setup and the capture photo function.

Variable Setup

```
int imageCount = 0;
String imageName = "image";
String imageEnding = ".jpg";
```

Capture Photo Function

```
void capture_photo(){
        String fileName = imageName + imageCount + imageEnding;
        char fileNameChar[50];
        fileName.toCharArray(fileNameChar,50);
        imageCount = imageCount +1;
```

```
// Check to see if the file exists:
// if it exists, delete the file:
if(sd.exists(fileNameChar)) sd.remove(fileNameChar);

// open a new empty file for write at end like the Native SD library
if (!myFile.open(fileNameChar, O_RDWR | O_CREAT | O_AT_END)) {
   sd.errorHalt("opening temp.jpg for write failed");
      }

// close the file:
myFile.close();

VC0706_compression_ratio(63);
delay(100);

VC0706_frame_control(3);
delay(10);

VC0706_frame_control(0);
delay(10);
rx_ready=false;
rx_counter=0;

Serial.end();                    // clear all rx buffer
delay(5);


Serial.begin(115200);

//get frame buffer length
VC0706_get_framebuffer_length(0);
delay(10);
buffer_read();

//while(1){};

// store frame buffer length for coming reading
frame_length=(VC0706_rx_buffer[5]<<8)+VC0706_rx_buffer[6];
frame_length=frame_length<<16;
frame_length=frame_length+(0x0ff00&(VC0706_rx_buffer[7]<<8))+VC0706_rx_
buffer[8];

vc_frame_address =READ_DATA_BLOCK_NO;

myFile.open(myFileChar, O_RDWR);
while(vc_frame_address<frame_length){
      VC0706_read_frame_buffer(vc_frame_address-READ_DATA_BLOCK_NO,
READ_DATA_BLOCK_NO);
      delay(9);

      //get the data with length=READ_DATA_BLOCK_NO bytes
```

```
        rx_ready=false;
        rx_counter=0;
        buffer_read();

        // write data to temp.jpg
        myFile.write(VC0706_rx_buffer+5,READ_DATA_BLOCK_NO);

        //read next READ_DATA_BLOCK_NO bytes from frame buffer
        vc_frame_address=vc_frame_address+READ_DATA_BLOCK_NO;

        }
// get the last data
vc_frame_address=vc_frame_address-READ_DATA_BLOCK_NO;

last_data_length=frame_length-vc_frame_address;


VC0706_read_frame_buffer(vc_frame_address,last_data_length);
delay(9);
//get the data
rx_ready=false;
rx_counter=0;
buffer_read();

myFile.write(VC0706_rx_buffer+5,last_data_length);

myFile.close();
}
```

CONCLUSION

This chapter provided you with a basic introduction to camera technology and an example module, which was easily usable with Arduino. We then extended the camera's basic functionality to turn it into a security device that records pictures whenever motion is detected.

This project has many opportunities for further expansion. For example, it can be integrated with almost any of this book's previous activities to periodically record images throughout the robot's motion.

However, since the Arduino is processing all the data from the camera, it is important to notice that it takes a considerable amount of time to capture the data, and during this time, the robot should not make any movements or perform any other functions. If you are interested in adding photo or video functionality into a more useful system, consider using a more powerful board such as the Arduino Dué or Arduino Mega.

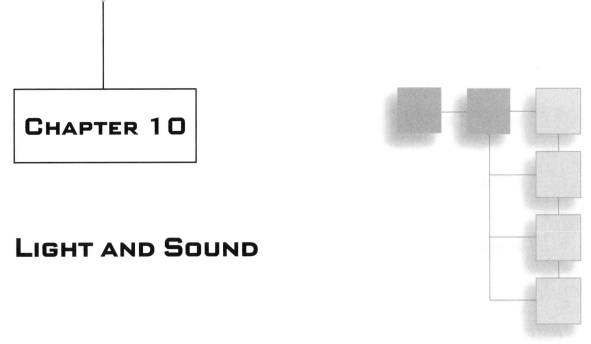

CHAPTER 10

LIGHT AND SOUND

In this chapter, you will learn how to integrate light and sound using a microphone sensor and a NeoPixel ring to generate an interesting light and sound show.

CHAPTER OBJECTIVES

- Learn how to program a NeoPixel array
- Learn how to use a microphone sensor
- Build a light and sound system with colored lights dancing to the beat of music

INTRODUCTION

This chapter continues the use of sensors to help you create a light and sound system in which multicolored lights dance to music. You will learn to program a NeoPixel array, which is an array of multicolored pixels. You will also learn how to use a sound sensor to detect the beats in music. You will then put the two together to create a multicolored array of lights that dance to your music.

MATERIALS REQUIRED

- Arduino board
- NeoPixel ring. A 12-LED ring from Adafruit.com: 12 x WS2812 5050 RGB LED ring.

- Sound sensor (microphone). One of the three sensors in the Robotics Sensor Kit from RadioShack. The kit also contains an optical sensor and an IR sensor.

- Solderless breadboard and hookup wires

This project is organized in three parts:

1. Hook up the NeoPixel ring to the Arduino board, and learn how to program the ring so that each pixel shows the color you want when you want.

2. Hook up the microphone to the Arduino board, and learn how the microphone gives a digital output depending on the noise level it detects.

3. Put it all together. Connect both the NeoPixel ring and microphone to the Arduino board and make the light dance to music.

PART 1: CONNECT THE NEOPIXEL RING TO THE ARDUINO AND PROGRAM PIXEL COLORS

The NeoPixel ring (model # 12 x WS2812 5050) is available from Adafruit.com. Adafruit makes several NeoPixel arrays. We used the 12-pixel ring because it is simple, provides excellent lighting effects, and can be programmed quite easily. A detailed description, as well as usage and programming details, are available at https://www.adafruit.com/products/1643. In this array, 12 RGB LEDs are assembled in a ring. The RGB stands for the three primary colors: red, green, and blue. Each pixel can be programmed to exhibit any color by appropriately mixing the primary colors via software.

Figures 10.1 and 10.2 show the front and back views of the NeoPixel ring.

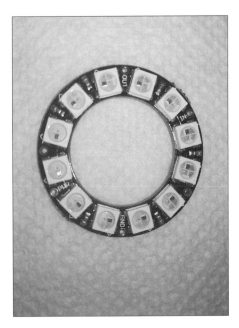

Figure 10.1
NeoPixel ring (front).
Source: Chandra Prayaga 2014

Figure 10.2
NeoPixel ring (back).
Source: Chandra Prayaga 2014

The ring has four terminals, to be connected as shown in Table 10.1.

Table 10.1 NeoPixel Ring Terminals	
NeoPixel Ring	**Arduino**
5V DC Power	%V
Ground (GND)	GND
Data Input	Digital pin 6
Data Out	No connection

Some soldering needs to be done for these connections and to get the ring ready for operation. The soldering steps are given below. For details, see: https://learn.adafruit.com/adafruit-neopixel-uberguide/best-practices.

1. Solder a capacitor (1000 µF, 6.3 V or higher) across the 5V DC and GND connections. We connected a 1000 µF, 35 V electrolytic capacitor (available at RadioShack), taking care to connect the – terminal of the capacitor to the GND on the ring, and the + terminal to the 5V on the ring.

2. Solder two leads—one to the 5V terminal on the ring, and the other to the GND terminal. The other ends of these wires can connect to either 5V on the Arduino and the GND on the Arduino, or to the + and – terminals of a 4.5 V independent battery pack, made of three 1.5 V batteries in series. If you use an independent battery pack, make sure to connect the – terminal of the battery pack to the GND on the Arduino.

3. Solder a 300 to 500 Ω resistor between the Digital pin 6 of the Arduino board and the Data Input connection on the ring. Using a hookup wire, connect one end to Arduino pin 6, and solder the other end to one end of the resistor. Solder the other end of the resistor to the Data Input connection on the ring. We used a 500 Ω resistor from RadioShack.

Figure 10.3 shows the capacitor soldered to the 5V DC and GND terminals on the ring. Figure 10.4 shows the resistor soldered to the Data In terminal, and the two leads for the 5V and GND leading to the battery pack, and also the lead to GND from the – terminal of the battery pack. Figure 10.5 shows the battery pack with the lead to the GND pin on the Arduino visible on the right.

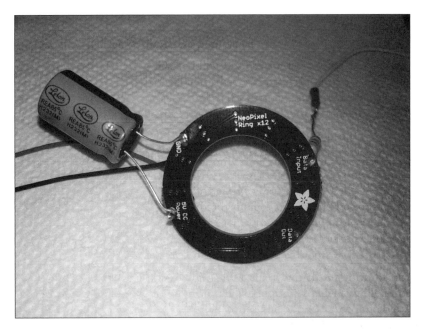

Figure 10.3
Capacitor.
Source: Chandra Prayaga 2014

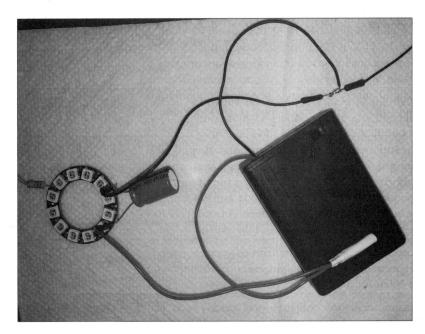

Figure 10.4
Resistor.
Source: Chandra Prayaga 2014

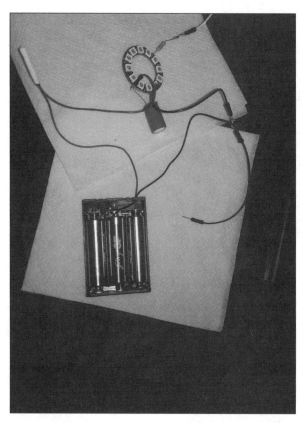

Figure 10.5
Battery pack.
Source: Chandra Prayaga 2014

Once the NeoPixel ring is ready for operation, we will plan a series of sketches, which will help us understand how to use it.

The NeoPixel ring comes with a library of commands. This library is available at the Adafruit website: https://learn.adafruit.com/adafruit-neopixel-uberguide/arduino-library>. Follow these instructions and install the library along with the other libraries that you installed. Programming the NeoPixel ring is easier with this library.

pixelColor0 Sketch

Our first question is: How can we light any one pixel with the color we want and then turn it off? The following sketch, pixelColor0, does this. It keeps blinking one particular pixel with one particular color. Run the sketch and experiment with the variables `pxlNum` and `pxlColor`. Give different values to these variables and see what happens.

```
// pixelColor0
// Switch on each pixel, one at a time, with one color, for desired time
#include <Adafruit_NeoPixel.h> // Library for all NeoPixel strips. We use the NeoPixel ring
//with 12 pixels.
#define PIN 6          //Arduino pin to send data to NeoPixel ring
Adafruit_NeoPixel strip = Adafruit_NeoPixel(12, PIN, NEO_GRB + NEO_KHZ800);
//Initialize an object called strip to represent the ring.
// The first argument is the number of pixels, in this case, 12. The second argument, PIN, is
//the Arduino pin to send data to the ring.
// The third argument specifies the wiring of the pixels (NEO_GRB) and the bitstream speed
//(NEO_KHZ800) for the ring.
// Other NeoPixel products need these arguments to be changed
// See https://learn.adafruit.com/adafruit-neopixel-uberguide/arduino-library for
//details

int pxlColor = strip.Color(0, 0, 255);   // Define integer variable called pxlColor.
//strip.Color() is a NeoPixel library function. The three arguments define RGB values, in
//this case, only Blue.
   // You can experiment with different values for the three colors. Each value must be
   //between 0 and 255. Each combination of the three values gives a different color.
   // The higher a given value, the brighter that particular color
int blank = strip.Color(0, 0, 0);     //Define integer for no color, called blank, all
//colors zero.
int pxlNum = 2;     // Define integer variable called pixelNum. This can be a value between 0
//and 11 to cover all twelve pixels. You should experiment by giving different values.

void setup()
{
  strip.begin();   // Prepare the data pin, in our case pin 6, for output to NeoPixel ring
  strip.show();    // Push data out to the ring. Since no colors have been set yet, this
                   //initializes all the NeoPixels to an initial "off" state.
}

void loop()
{
  strip.setPixelColor(pxlNum, pxlColor);   //Library function, sets the color of pixel
     //pxlNum to the value pxlColor. In our case, we set pxlNum to 2 and pxlColor to Blue when
     //we declared these variables.
  strip.show();   // Library function, push the data out to the ring. This puts the pixel
                  // pxlNum on with color pxlColor.
  delay(500);     // Leave it on for 1/2 second
  strip.setPixelColor(pxlNum, blank);   //Set that pixel to off
  strip.show();   // Push that data out to the ring
}
```

How the Code Works

This pixelColor0 code illustrates how to pick a specific pixel and set it on and off. The following is how this code works.

Include a library for the NeoPixel strip. The one used in this code is the NeoPixel ring.

Pin 6 is used to send data to the NeoPixel ring.

Initialize an object called strip to represent the ring.

The first argument is the number of pixels, in this case, 12. The second argument, PIN, is the Arduino pin to send data to the ring. The third argument specifies the wiring of the pixels (NEO_GRB) and the bitstream speed (NEO_KHZ800) for the ring.

Other NeoPixel products need these arguments to be changed.

See https://learn.adafruit.com/adafruit-neopixel-uberguide/arduino-library for details.

Define an integer variable called pxlColor. strip.Color() is a NeoPixel library function. The three arguments for the strip.Color() function define RGB values. In this case, only Blue.

You can experiment with different values for the three colors. Each value must be between 0 and 255. Each combination of the three values gives a different color.

The higher a given value, the brighter that particular color.

Define an integer for no color, called blank, which sets all colors to zero.

Define an integer variable called pixelNum. This can be a value between 0 and 11, to cover all twelve pixels. You should experiment by giving different values.

The setup() function does the following:

The data pin 6 is prepared to send output to the NeoPixel ring.

The data is pushed to the pixel. Since no colors have been set, all pins on the strip are initialized to an off state.

The loop() function does the following:

The library function strip.setPixelColor(pxNum, pxColor) sets the color of pixel pxlNum to the value pxlColor.

In our case, we set pxlNum to 2 and pxlColor to Blue when we declared these variables.

strip.show(), a library function, pushes the data out to the ring. This puts the pixel pxlNum on with color pxlColor.

The color remains on for 1/2 second per the delay and then sets that pixel to an off state. Lastly, it pushes that data out to the ring.

pixelColor1 Sketch

Once you get used to this simple idea of sending commands to the NeoPixel ring, which can turn on any pixel with any color, you can try out sketches that are more complicated and do some fascinating operations with the ring. For example, the following sketch, pixelColor1, turns on the entire ring of 12 pixels with a specified color. Note the for() loop that runs through the entire ring.

```
// pixelColor1
// Getting used to a NeoPixel ring.
// This program switches on and, after a desired delay, switches off each pixel, one after
// the other, with one color of your choice, and repeats the cycle, going round the ring.

#include <Adafruit_NeoPixel.h> // Library for all NeoPixel strips. We use the NeoPixel ring
                                //with 12 pixels.

#define PIN 6            // Arduino pin to send data to NeoPixel ring

Adafruit_NeoPixel strip = Adafruit_NeoPixel(12, PIN, NEO_GRB + NEO_KHZ800);
// Initialize an object called strip to represent the ring.
// The first argument is the number of pixels, in this case, 12. The second argument, PIN, is
//the Arduino pin to send data to the ring.
// The third argument specifies the wiring of the pixels (NEO_GRB) and the bitstream speed
//(NEO_KHZ800) for the ring.
// Other NeoPixel products need these arguments to be changed
// See https://learn.adafruit.com/adafruit-neopixel-uberguide/arduino-library for
//details

void setup()
{
  strip.begin();    // Prepare the data pin, in our case pin 6, for NeoPixel output
  strip.show();     // Push data out to the pixel. Since no colors have been set yet, this
                    // initializes all the NeoPixels to an initial "off" state.
}

void loop()
{
  int pxlColor = strip.Color(0, 0, 255); // strip.Color() is a NeoPixel library function.
    //The three arguments define RGB values. In this case, only Blue.
```

```
    // You can experiment with different values for the three colors. Each value must be
    //between 0 and 255. Each combination of the three values gives a different color
    // The higher a given value, the brighter that particular color
    colorShow(pxlColor, 500); // Our function, defined below, specifies the pixel color and
    //the delay time for which it should be on. In this case, Blue, 1/2 second delay
    delay(100);
}
void colorShow(int c, int wait)
{
    int blank = strip.Color(0, 0, 0);      //Blank, all colors zero.
    for(int i = 0; i < strip.numPixels(); i++)      //strip.numPixels() is a library function
        //that returns the number of pixels in the array, in our case, 12
    {
        strip.setPixelColor(i, c);      // Library function, sets the color of pixel i to the
            //value c, passed to the function. In our case, Blue.
        strip.show();      // Library function, push the data out to the ring. This puts the
                           //pixel i on.
        delay(wait);       // delay, in this case, 1/2 second
        strip.setPixelColor(i, blank); // After the delay, shut the pixel off
    }      // This for loop runs round the ring, blinking each pixel on and off, then passing on
        //to the next pixel
}
```

How the Code Works

This code switches on each pixel on the LED strip one at a time for a desired time. The following is how this code works.

Include a library for the NeoPixel strips. The one used in this code is the NeoPixel ring.

Pin 6 is used to send data to the NeoPixel ring.

Initialize an object called `strip` to represent the ring.

The first argument is the number of pixels, in this case, 12. The second argument, PIN, is the Arduino pin to send data to the ring. The third argument specifies the wiring of the pixels (NEO_GRB) and the bitstream speed (NEO_KHZ800) for the ring.

Other NeoPixel products need these arguments to be changed.

See https://learn.adafruit.com/adafruit-neopixel-uberguide/arduino-library for details.

The `setup()` function does the following:

The data pin 6 is prepared to send output to the NeoPixel ring.

The data is pushed to the pixel. Since no colors have been set, all pins on the strip are initialized to an off state.

The `loop()` function does the following:

`strip.Color()` is a NeoPixel library function. The three arguments define RGB values. In this case, only Blue. You can experiment with different values for the three colors. Each value must be between 0 and 255. Each combination of the three values gives a different color. The higher a given value, the brighter that particular color

Our function, `colorShow()`, defined below, is called with a specific color and a set delay, in this case, Blue, with a 1/2 second delay.

The `colorShow()` function does the following:

Define a variable called `blank`, which sets all the RGB color values to zeros. This will be used to shut each pixel down.

The `for()` loop runs round the ring, one pixel at a time. In the loop, each pixel is turned on with the color passed to the function. The library function `strip.show()` pushes the data out to the ring and puts the pixel on.

After the specified delay, the pixel is turned off by setting the color to `blank` and then calling `strip.show()` again.

The loop then goes to the next pixel.

pixelColor2 Sketch

In the following sketch, pixelColor2, another feature is added. The entire ring is turned on and off, but each time the color of the pixels is changed randomly. You will also see the use of a random number generator.

Here is the sketch:

```
//pixelColor2
//Program to randomly change the color of pixels. Turn the entire ring on with a random
//color, and then turn it off after a delay.
#include <Adafruit_NeoPixel.h>
#define PIN 6

Adafruit_NeoPixel strip = Adafruit_NeoPixel(12, PIN, NEO_GRB + NEO_KHZ800);


int r;    // Random integer to change color
int c;    // Color value
```

```
void setup()
{
  strip.begin();
  strip.show(); // Initialize all pixels to 'off'
}

void loop()
{
  randomSeed(analogRead(0));   // Standard Arduino function to seed a random number
                               // generator
  r = random(3); // Select a random number with possible values 0, 1, 2. Depending on this
                 // value, the pixels will have a different color.
  onPixels(r);   // Function defined below to turn on all pixels with a color defined by the
                 // value of r.
  delay(500);    // Keep the pixels on for this time
  offPixels();   // Our function defined below to turn off all pixels.
}

void onPixels(int q)      // q is the random number generated each time
{
  if (q == 0)                    //If the random number is 0
  {
    c = strip.Color(250, 0, 0);      // Color for all pixels will be red
  }
  else if (q == 1)      // If the random number is 1
  {
    c = strip.Color(0, 250, 0);      // Color of all pixels set to green
  }
  else
  {
    c = strip.Color(0, 0, 250);      // If the random number is neither 0 nor 1 (it is 2), set
                                     //the pixels' color to blue
  }
  for (int p = 0; p < 12; p++)    // Go round the ring
  {
    strip.setPixelColor(p, c);      // Set each pixel to the color randomly picked above
  }
  strip.show();       // Send the data to the ring via pin 6
}

void offPixels()
{
  for (int p = 0; p < 12; p++)    // Go round the ring
```

```
  {
    strip.setPixelColor(p, strip.Color(0, 0, 0));      // Set each pixel color to blank to
                                                       //shut it down
  }
  strip.show();    // Send the signal to the ring and shut down the entire ring
}
```

How the Code Works

This code randomly allocates a color to the pixels in the pixel strip. Following is how the code works.

Include a library for the NeoPixel strips. The one used in this code is the NeoPixel ring.

Pin 6 is used to send data to the NeoPixel ring.

Initialize an object called strip to represent the ring.

The first argument is the number of pixels, in this case, 12. The second argument, PIN, is the Arduino pin to send data to the ring. The third argument specifies the wiring of the pixels (NEO_GRB) and the bitstream speed (NEO_KHZ800) for the ring.

Other NeoPixel products need these arguments to be changed.

See https://learn.adafruit.com/adafruit-neopixel-uberguide/arduino-library for details.

Define a variable called r, which will provide a random integer to change the color of the pixel.

Define an integer variable called c, to hold the color value.

The setup() function does the following:

The data pin 6 is prepared to send output to the NeoPixel ring.

The data is pushed to the pixel. Since no colors have been set, all pins on the strip are initialized to an off state.

The loop() function does the following:

A standard Arduino function is used to seed a random number generator.

Select a random number with possible values 0, 1, or 2. Depending on this value, the pixels will have a different color.

Call the onPixels(r) function defined below to turn on all pixels with a color defined by the value of r.

Keep the pixels on for the provided delay of 1/2 a second.

Call the offPixels() function defined below to turn off all pixels.

The onPixels(int q) function does the following:

q is the random number generated each time. If the random number is 0, color for all pixels will be red; if the random number is 1, color of all pixels is set to green. If the random number is neither 0 nor 1 (it is 2), set the pixels' color to blue.

Set all pixels to the color randomly picked above. Send the data to the ring via pin 6.

The offPixels() function does the following:

Set all pixel colors to blank to shut them down. Send the signal to the ring.

The three sketches above showed you the basics of how to use the NeoPixel ring. Obviously, you can think of many other possibilities. For example, you could have some of the pixels show one color, and some other pixels show a different color, and even distribute a rainbow effect. The sample sketch that comes with the library, called "strandtest," shows some more examples, including the rainbow effect. When you installed the Neo-Pixel library, you also installed this sketch. In the IDE, click on File > Sketchbook > Libraries > Adafruit_NeoPixel > strandtest.

PART 2: ATTACHING THE MICROPHONE

Now that you've learned how to use the NeoPixel ring, you can start the process of coupling it with a microphone sensor so that you can combine sound with light. In this part, you'll attach the microphone to the Arduino board and learn how the microphone gives a digital output depending on the noise level it detects. The microphone sensor is one of three sensors in the Robotics Sensor Kit from RadioShack. The kit also contains an optical sensor and an IR sensor. We will use the microphone sensor.

The microphone sensor is a packaged tiny microphone with three connector pins attached to the package. The pins are labeled S (Signal), G (GND) and V (5 V). Figure 10.6 shows a picture of the sensor package.

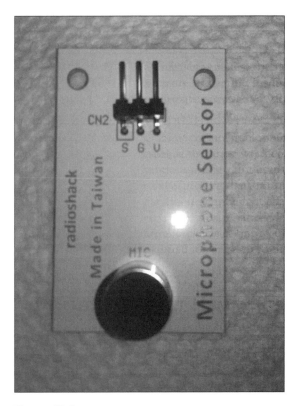

Figure 10.6
Microphone sensor package.
Source: Chandra Prayaga 2014

The microphone is powered by connecting the G and V pins to GND and 5V, respectively, on the Arduino, and the signal pin to any of the digital pins. In our setup, we connected the signal pin to pin 7 on the Arduino. The microphone sensor gives out a digital signal of 0 (LO) whenever it hears a loud sound, and 1 (HIGH) when there is low or no sound. The connections for the setup of the microphone are shown in Table 10.2.

Table 10.2   Microphone Connections

Sensor Pin	Arduino Pin
S	7
G	GND
V	5V

soundSensor1

The sketch soundSensor1 provided below illustrates how to use the microphone sensor.

```
// Program to control and use the RadioShack microphone sensor


int pin = 7;      //Digital output from the microphone connected to pin 7 on Arduino
int val;          // Variable to hold the digital value from pin 7
// Digital output from microphone sensor is HIGH if the sound is low in volume, and LO if the
//sound is high in volume
// Use the Serial Monitor to check the output
void setup()
{
  Serial.begin(9600);      // Ready the Serial Monitor
  pinMode(pin, INPUT);     // Configure pin 7 for reading data input to Arduino from the
                           //microphone
}
void loop()
{
  val = digitalRead(pin);   // Assign digital value of microphone output to the variable val
  if(val ==0)               // If there is a loud sound
  {
    Serial.println("Woot");   //Acknowledge loud sound detected by microphone
    delay(500);   // Delay 1/2 second before checking output again
  }
}
```

How the Code Works

This code illustrates how the microphone sensor from RadioShack works. The following describes the process of this code.

Digital output from the microphone is connected to pin 7 on the Arduino board. val is a variable to hold the digital value from pin 7.

The setup() function does the following:

Start the serial monitor.

Configure pin 7 for reading data input to the Arduino from the microphone.

The `loop()` function does the following:

Assign digital value of microphone output to the variable `val`.

If `val` equals `0`, acknowledge loud output from microphone. Delay 1/2 second before checking output again.

Once you have connected the microphone as described above, clap your hands in front of the microphone. Each time you clap, you will see a "Woot" displayed on the Serial monitor, which, you will remember, is opened by pressing Ctrl + shift + N on the keyboard as soon as you start running the sketch.

PART 3: PUTTING IT ALL TOGETHER

Now you can connect both the NeoPixel ring and the microphone sensor to the Arduino board and make the light dance to music. Set up both by connecting them to the Arduino, and design a sketch so that whenever there is a loud sound, it triggers a NeoPixel ring reaction, with whatever colors you want. If there is music playing, then for each loud beat in the music, you can make the lights dance and change color. Figure 10.7 shows the setup with both the NeoPixel ring and the microphone sensor connected to the Arduino as described above. The connections are as described in both Tables 10.1 and 10.2.

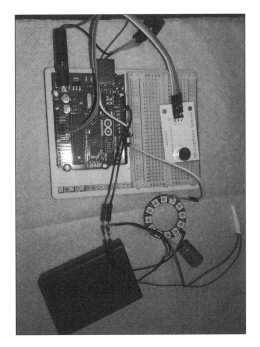

Figure 10.7
NeoPixel ring and microphone sensor connected to the Arduino.
Source: Chandra Prayaga 2014

In the sketch, soundLight4, watch for the condition that triggers the NeoPixel ring to light up whenever a loud sound is heard by the microphone, as shown by a value of 0 at pin 7 of the Arduino, which is connected to the microphone sensor output.

soundLight4

```
// Light and sound display. NeoPixel ring blinks and changes colors to the beat of music.

#include <Adafruit_NeoPixel.h>    // NeoPixel library
#define PIN 6      // Arduino pin for data input to ring

Adafruit_NeoPixel strip = Adafruit_NeoPixel(12, PIN, NEO_GRB + NEO_KHZ800);
//Initialize object called strip, of type Adafruit_NeoPixel. This object represents our Ring.


int pin = 7;    // Arduino pin connected to output of microphone
int val;        // Variable to hold value of microphone output
int r;        // Random variable to hold a value 0, 1, or 2
int c;        // Variable to hold value of color of pixels
void setup()
{
  pinMode(pin, INPUT);        // Arduino pin 7 reads microphone output
  strip.begin();
  strip.show();  //  Initialize all pixels to 'off'
  randomSeed(analogRead(0));      // Random variable needs to be seeded
}
void loop()
{
  val = digitalRead(pin);      //Read digital value at pin 7 and assign to val
  if(val ==0)      // If the microphone senses a loud sound
  {
    r = random(3);        // Assign value of random variable of 0, 1 or 2 to r
    onPixels(r);      // Call function to turn on pixels with a random color determined by
                      //the value of variable r
    delay(100);
  }
  else
  {
    offPixels(); // If the sound is low, call function to turn off the pixels
  }
}

void onPixels(int q)      // Define function to turn pixels on. Input parameter to the
//function is an integer. We call this function above with random variable r as input
```

```
{
  if (q == 0)       // If random variable is zero
  {
    c = strip.Color(200, 0, 0);    // Set color variable to RED
  }
  else if (q == 1)   // If random variable is 1
  {
    c = strip.Color(0, 200, 0);    // Set color variable to GREEN
  }
  else               // If random variable is 2
  {
    c = strip.Color(0, 0, 200);    // Set color variable to BLUE
  }
  for (int p = 0; p < 12; p++)     // For all the pixels
  {
    strip.setPixelColor(p, c);      // Set pixel color to random color given by variable c
  }
  strip.show();      // Send the data to the ring
}
void offPixels()     // Define function to turn off all pixels
{
  for (int p = 0; p < 12; p++)   //For all the pixels
  {
    strip.setPixelColor(p, strip.Color(0, 0, 0));     // Set the color to blank, or off
  }
  strip.show();      // Send the data to the ring
}
```

How the Code Works

This code is a light and sound display. The NeoPixel ring blinks and changes colors to the beat of music. Here's how the code works:

Include the NeoPixel library.

Pin 6 is used to send data to the NeoPixel ring.

Initialize an object called strip, of type Adafruit_NeoPixel. This object represents our ring.

Arduino pin 7 is connected to the output of the microphone.

val is a variable to hold the value of the microphone output.

r is an integer variable to hold a random value: 0, 1, or 2.

c is an integer variable to hold the value of the color of pixels.

The setup() function does the following:

Arduino pin 7 reads the microphone output.

The data pin 6 is prepared to send output to the NeoPixel ring.

The data is pushed to the pixel. Since no colors have been set, all pins on the strip are initialized to an off state.

A random variable gets seeded.

The loop() function does the following:

Select a random number with possible values 0, 1, or 2. Depending on this value, the pixels will have a different color.

Call the onPixels(r) function defined below to turn on all pixels with a color defined by the value of r.

Keep the pixels on for the provided delay of 1/2 a second

Call the offPixels() function defined below to turn all pixels off.

The onPixels(int q) function does the following:

q is the random number generated each time.

If the random number is 0, the color for all pixels will be red.

If the random number is 1, the color of all pixels is set to green.

If the random number is neither 0 nor 1 (it is 2), set the pixels' color to blue.

The for() loop sets all pixels to the color randomly picked above.

Send the data to the ring via pin 6.

The offPixels() function does the following:

Set all pixel colors to blank to shut them down.

Send the signal to the ring.

CONCLUSION

In this chapter, you have seen how to combine different elements such as sound and light to design interesting applications. You have specifically used the NeoPixel ring sensor and the microphone sensor to design a sound and light show. You can extend this application to set up a Christmas tree light and sound show or simply use it as an addition to a party.

CHAPTER 11

ANDROID APP CONTROLLER

Previously, all robotics activities required a computer present to send commands or retrieve data. In today's world, mobile phones have created an environment free of physically tethered devices. This chapter explores the development of a mobile app on the Android OS, which can remotely control your robot!

CHAPTER OBJECTIVES

- Install and learn the Netbeans IDE
- Learn basics of the Java programming language
- Connect to your Arduino via Java
- Create a graphical Java application using JavaFX
- Remotely control your robot

Note

This section requires that you have completed Chapter 8, "Remote-Controlled User Interfaces," creating a server application on Arduino and connecting to it via Java. You will only be developing the Android app in this section, reusing all of your Arduino code from Chapter 8.

INTRODUCTION

Mobile phones, and the services that run them, have made a defining change in the technological revolution. Today, more than 90% of American adults have a cell phone; 52% of whom have smart phones; while 34% of these smart phone users use their phone as their primary Internet device! With such a proliferation of mobile technologies, other features of mobile phones have changed with the times, with features such as: Wi-Fi, Bluetooth, NFC (near-field communication), enterprise security, and just about anything you can find on the average computer. These features, packed in a device that's always in our pocket, yield some life changing applications, including the ability for remote control of other devices…like an Arduino! In this chapter, you will use your existing server application, which you created back in Chapter 8, and create an Android application that can be run on your phone (or emulator) and used to control your robot over Wi-Fi.

MATERIALS REQUIRED

- Standard Arduino robot used in previous chapters
- Arduino Wi-Fi shield
- USB mini cable
- Android phone (or free Android emulator)
- Wi-Fi network
- Computer with Internet access
- Android Developer Kit (free)

PART 1: GETTING READY FOR ANDROID

We'll begin with an introduction of Android app development, including the differences between Android apps and programs you have written throughout the book, and then move into the required hardware and software tutorials to begin development. At the end of this section, you will be able to download and compile an application to an Android phone or emulator. In Part 2, we will develop an app to specifically control your Arduino.

Android Programming Architecture and Language

Android applications, like all mobile applications, must be thought of differently than traditional embedded, desktop, or web applications. For example, we are limited by the amount of screen space we have available to us; users expect to interact with the app through touch versus a more controlled mouse; and while mobile devices are quite fast, we must optimize our applications to be quick on smaller devices. Most apps adapt to these challenges through a methodology called Model-View-Controller (MVC), which describes the three distinct components of an app (see Figure 11.1).

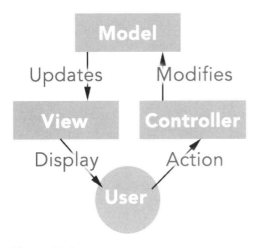

Figure 11.1
MVC diagram.

The Model contains the "state" of the application and contains all the data used by the app. This data, perhaps user information, is requested by the View component, which presents this data to the user in the form of a screen. A View is typically a layout that populates with data, images, and external references provided by the model; it is what the user sees. The Controller component "listens" to what the user does to the View (e.g., swipe left, click picture, etc.). The Controller then calls the Model to change or perform a calculation based on the user's input. This change in the Model triggers an update to the View, and the user then has another chance to interact with it.

This methodology, or architectural pattern, provides a clear separation between the graphical design aspects (front-end) of apps, mostly done in the View, from the more computationally heavy actions (back-end) done in the Controller and Model. While the programmer has the ultimate decision as to where each function is located, Software Development Kits (SDKs) like Android and iOS, as well as frameworks (which are used for more advanced apps), are based on the MVC architecture. Android packages these three components into an "Activity," and multiple Activities make up an application. Keep this in mind as we progress through creating an Android app.

Android is a framework of libraries that runs on a specialized Java virtual machine (JVM). Therefore, any legal Java software can run on Android; however, unless you use the special libraries, you will not be able to do much. The Android libraries provide the abilities to access all phone components, including the screen, buttons, microphone, etc. Android also uses a special pre-compiler, which helps to keep your project clean and organized. Much of the time, this pre-compiler simply creates lists of image assets you include, although more functions are used in more advanced projects. With these concepts in mind, let's download and install the appropriate software to begin exploring Arduino.

Installing the Android Developer Kit

Android provides a variety of ways to start developing applications; however, the simplest is with the Android Developer Kit, a specialized Eclipse IDE. This software will set up everything you need to create a sample project on Arduino.

1. To download this software, use the following link: http://developer.android.com/sdk/index.html.

2. Click on the Download Eclipse ADT button, accept the Terms and Conditions, and then select Download. Move the downloaded ZIP file to a useful location (such as "Applications" or a development folder) and unzip the file.

3. Within the unpacked folder are two sub-directories: "eclipse" and "sdk." sdk contains all the required files for the Arduino libraries, while eclipse contains the development environment in which you will create your applications. Open the "eclipse" folder and open the eclipse application. Your screen should look like Figure 11.2. If you are asked to select a workspace, the default location is fine; just click OK.

Figure 11.2
Blank ADT interface.
Source: Eclipse

4. You will likely be asked to update the Android SDK. Select Install and ensure that Android 4.4.2 is selected, as in Figure 11.3. Click Install Packages at the bottom right.

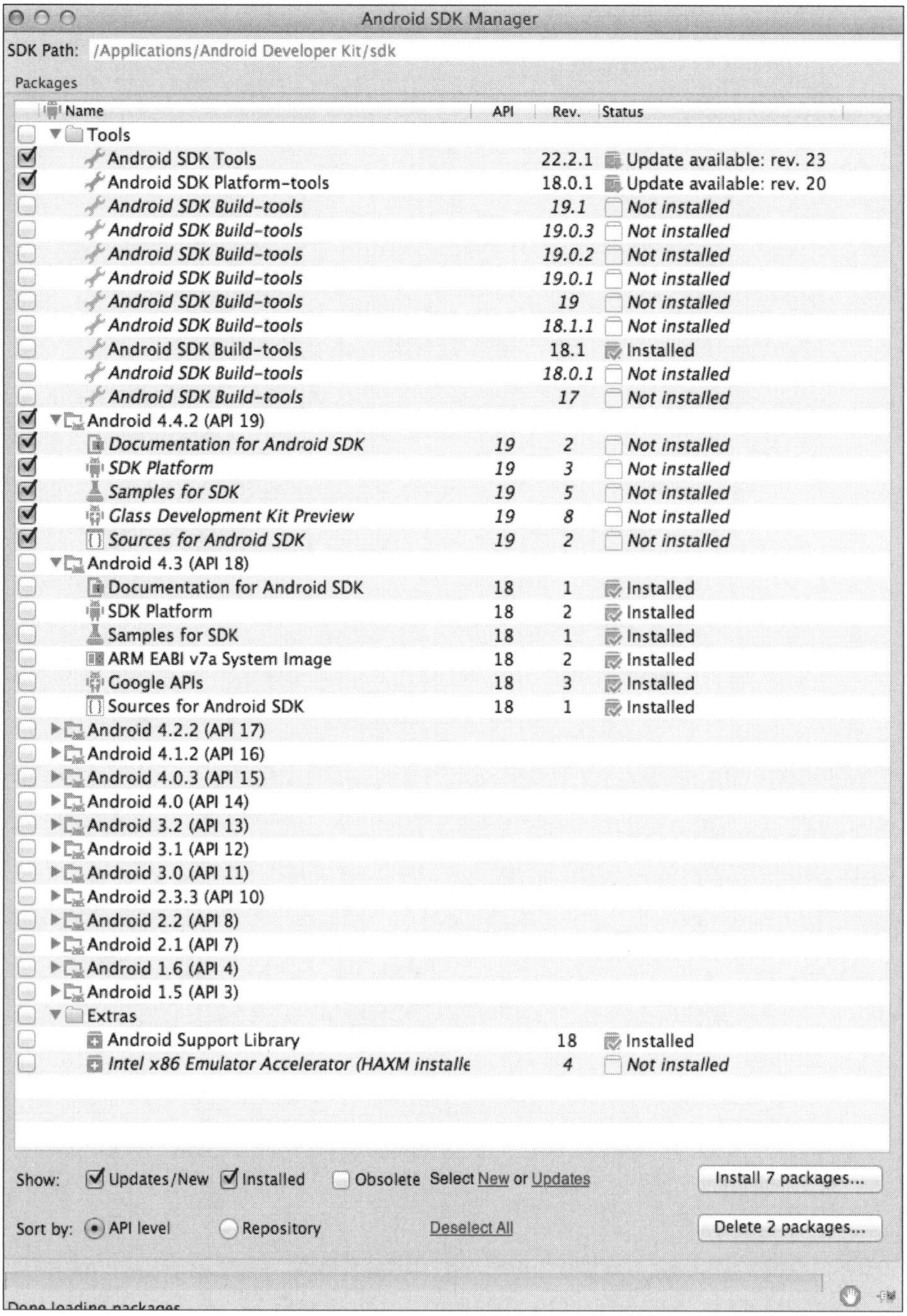

Figure 11.3
Android SDK package updater.
Source: Arduino

5. When prompted, select each package on the left, and accept the terms on the bottom right, as shown in Figure 11.4.

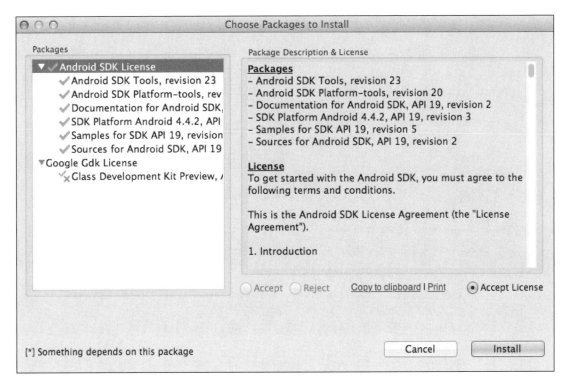

Figure 11.4
Accept Android updater terms.
Source: Arduino

Installing the Genymotion Android Emulator

The Genymotion emulator allows you to test your applications on your computer on a virtual Android device, and will make testing easier, whether or not you have an Android device.

1. Open Genymotion using the following link: https://cloud.genymotion.com/page/launchpad/download/

2. Create an account using the "signup form" link and follow the instructions to create an account. Once activated, return to the Genymotion link and download the free version by clicking Download.

3. Select your platform and download the appropriate file. If you are on a Windows machine, VirtualBox is included and will be installed automatically. If you are running Mac or Linux, download VirtualBox from this link: https://www.virtualbox.org/wiki/Downloads

4. If you are on Mac or Linux, follow the VirtualBox installation instructions. If you installed it correctly, you should be able to open VirtualBox and see a window similar to that shown in Figure 11.5.

Figure 11.5
VirtualBox interface.
Source: Oracle VirtualBox

5. Start the Genymotion installer, and follow the instructions to complete the installation. Open the Genymotion application, which should look like Figure 11.6.

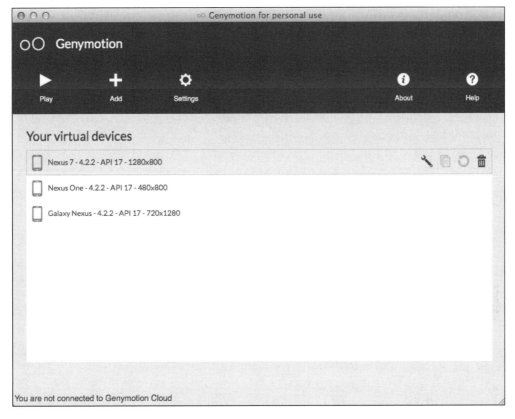

Figure 11.6
Genymotion interface.

Source: Genymobile

6. Select the Add button to create a new virtual device. Log in using the account you made previously, and you should see a list of available virtual devices (Figure 11.7).

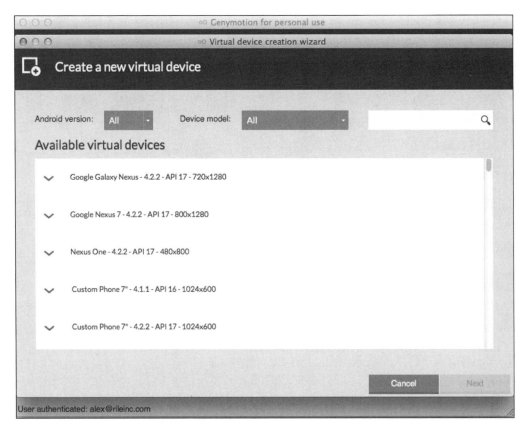

Figure 11.7
Add device interface.
Source: Genymobile

7. Select the "Google Nexus 7 - 4.2.2 - API 17 - 800×1280"odevice and click Next. Wait for the device to be installed and then click Finish. You should now see that device in your Genymotion interface. Double-click it to launch the virtual device. After booting, you will see an Android virtual machine, as shown in Figure 11.8.

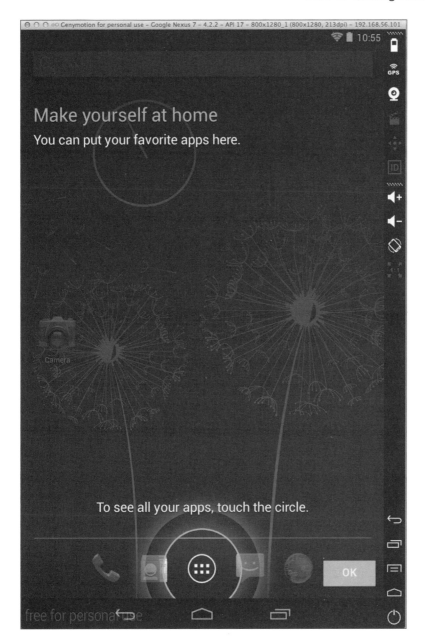

Figure 11.8
Android virtual machine.

Source: Genymobile/Android

8. This virtual machine is in fact running Android, and can be used just like a physical Android phone. Minimize this application for now; we'll be using it for testing soon.

Creating a Sample Application

1. Return to the Eclipse workspace and right-click in the white area under the Package Explorer tab. Select New, and then select New Android Application.

2. Enter "Arduino Controller" into the Application Name. Change the Target SDK to "API 19: Android 4.4". Select Next.

3. Select Next until prompted to create an Activity. Select Blank Activity and click Finish. Your Eclipse IDE should populate like Figure 11.9.

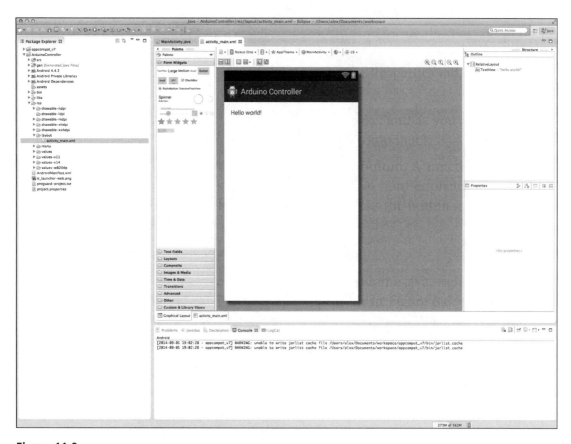

Figure 11.9
Eclipse IDE.
Source: Eclipse

4. To run the sample application, we need to create a "Run Configuration." In the top menu, select Run, and then Run Configurations. Double-click Android Application to create a new configuration. Edit the configuration by renaming it "Default Run" and selecting ArduinoController in the Project field (Figure 11.10).

Figure 11.10
Creating the default run configuration.
Source: Eclipse

5. Apply the changes and click Run. You will be asked to choose a running Android device. You should see the genymotion-google_nexus device we spun up earlier (Figure 11.11). Select it and click OK. If you do not see the device, you can re-open Genymotion and double-click the device again.

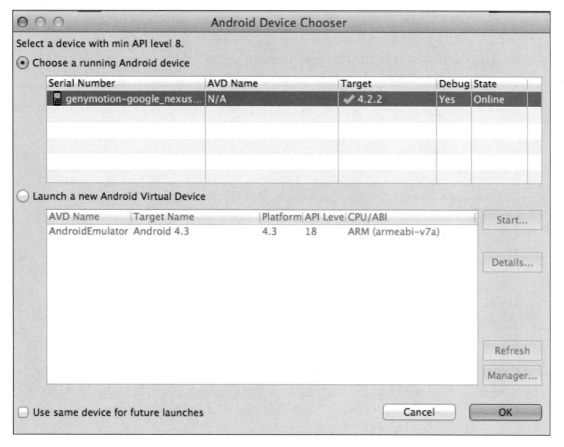

Figure 11.11
Choosing a device.
Source: Eclipse

6. The emulator will load the app, and you should see a screen like Figure 11.12.
   Congratulations, you just created your first app!

Figure 11.12
Default Hello World! application.
Source: Genymobile/Android

7. You can also use your own Android device to run your application. If you have an Android device, you must first enable USB Debugging. This is dependent on the physical device you have, and a quick Internet search can give you instructions. For example, on the Droid 2 running Android 2.4.2, USB Debugging is enabled in the Settings > Applications > Development menu (Figure 11.13).

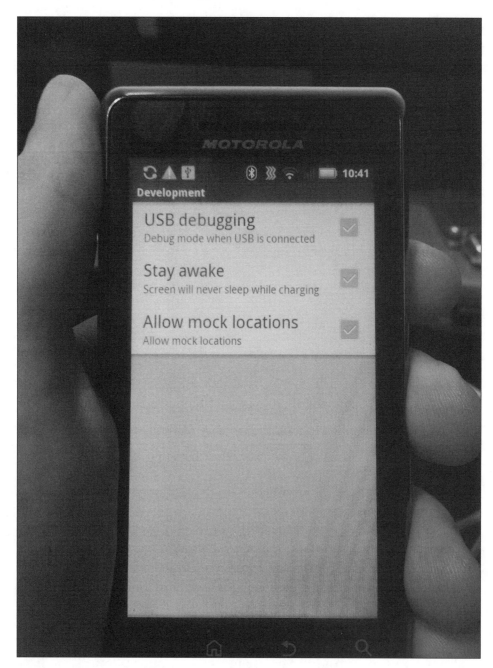

Figure 11.13
Development settings on Android 2.4.2.

8. Once USB Debugging is enabled, your phone will appear as an option whenever you run your application (Figure 11.14). Simply select it and click Run.

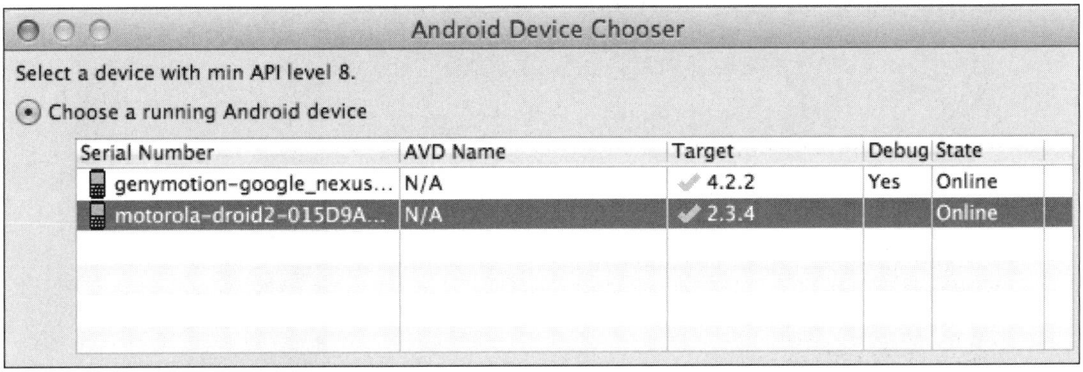

Android Device Chooser				
Select a device with min API level 8.				
⊙ Choose a running Android device				
Serial Number	AVD Name	Target	Debug	State
📱 genymotion-google_nexus...	N/A	✔ 4.2.2	Yes	Online
📱 motorola-droid2-015D9A...	N/A	✔ 2.3.4		Online

Figure 11.14
Selecting your device for debugging.
Source: Eclipse

You have now installed all you need to create Android apps and run them on either the Genymotion emulator or your physical phone!

PART 2: CREATING AN ARDUINO CONTROLLER APP

In this section, we'll start by going into more detail of each of the files within your app, and how to connect them to your Arduino, just as you connected your Swing application in Chapter 8.

You should have the activity_main.xml file open to start the project. It is located under res/layout/activity_main.xml. You should see an Android screen with tool palettes along the edges. In this layout (Figure 11.15), you can add components to your View, which is displayed to the user.

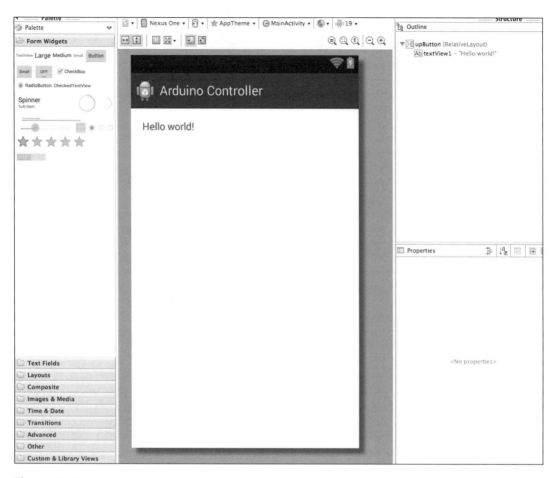

Figure 11.15
Android XML editor.
Source: Eclipse

1. Start by editing the "Hello World" text box. Click once on "Hello World" to select the box. Its properties will populate the Properties box at the bottom right of the interface. In the Text property, double-click on the value and enter "Hello, use this interface to control your Arduino!" You will see the screen update to display your text.

Note

If you double-click on the screen, it will display the original XML document. This is how the computer sees your screen, and what it uses to render your screen. Simply click on the Graphical Layout tab to return to the button view (Figure 11.16).

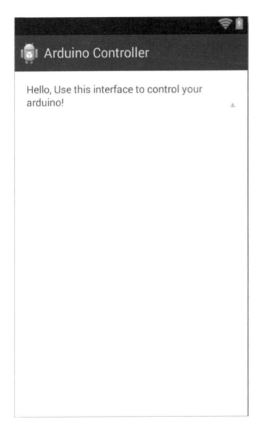

Figure 11.16
Updated welcome screen.
Source: Genymobile/Android

2. You will also need to create buttons to move your robot. In the top left corner of the IDE is a tab called Form Widgets, and inside is a "Button" button. Drag five of them to the screen in an arrow pad (up/down/left/right/center) configuration (Figure 11.17).

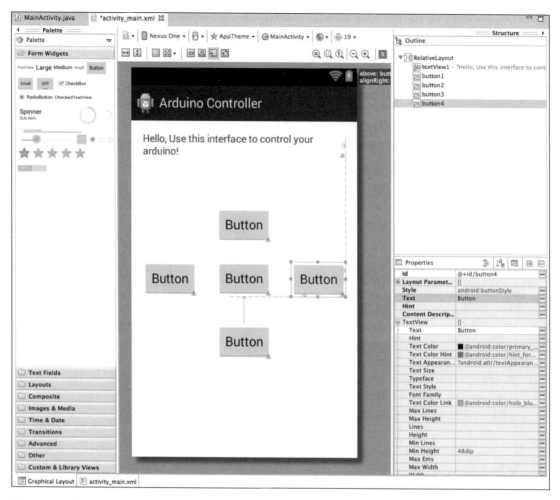

Figure 11.17
Buttons on user interface.
Source: Eclipse

3. Rename each of your buttons by clicking them and editing their Text property. This changes how they appear on the screen.

4. In order to reference these buttons in your code, you will need to provide them with meaningful identifiers. Click on the Up button. The top property should be the Id property. This is its Android ID, which represents how it is referenced throughout the app. To change it, click in the Id's value and enter a new ID, such as upButton. The IDE will ask if you want to replace the identifier. Select Yes.

Repeat this process for each button with its corresponding ID (e.g., downButton, leftButton, rightButton, stopButton). You have now created an appropriate View with which the user may interact.

5. We will now create a Controller to handle the user's actions, and Model components to connect to the Arduino. Open the MainActivity.java file. It is located under src/com.example.arduinocontroller.

We first need to declare the variables we'll need to connect to and read from/write to the Ardunio. Directly below the class declaration

```
public class MainActivity extends ActionBarActivity {
```

add three variables: a `Socket`, an `InputStream` and an `OutputStream`.

```
public class MainActivity extends ActionBarActivity {
        Socket soc;
        InputStream in;
        OutputStream out;
```

It is likely that you will be given an error for each of the declarations in the above code snippet. Simply click on the light bulb and select import.

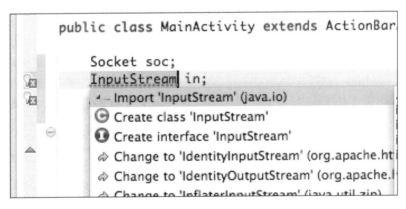

Figure 11.18
Importing unresolved libraries in Eclipse.
Source: Eclipse

6. We will now create a connect method to connect to the Arduino. You can create this method just below the variable declarations.

```
Socket soc;
InputStream in;
OutputStream out;

public void connect()
{


}
```

Start the method with a try{}catch(){} loop. Connecting to a socket is an uncertain event for the computer (the other computer may not exist), so the Java language requires that you explicitly handle problems, called exceptions.

```
public void connect()
{
    try{


    }catch(Exception e)
    {

    }
}
```

We'll place our connection code inside the try block, creating a new socket connection, and then initiating input and output streams. You'll need to know your Arduino's IP address (see Chapter 8). In the example below, my Arduino's IP address is 192.168.1.60. Replace this number with your Arduino's IP address. The 23 represents the port in which we created the Arduino server back in Chapter 8.

```
public void connect()
{
    try{
      soc = new Socket("192.168.1.60",23);
      in = soc.getInputStream();
      out = soc.getOutputStream();
    }catch(Exception e)
    {

    }
}
```

7. We'll now handle the error that may be generated from the connection code. In this case, it is best to alert the user through a status message that the Arduino is unreachable. Inside the `catch` block, we'll create an `AlertDialog`, which is an Android object for displaying errors.

```
}catch(Exception e)
        {
            new AlertDialog.Builder(this)
        }
    }
```

You'll notice that we did not put a semicolon at the end of this code, because we can modify this object without creating a variable by simply using the dot notation with which you are familiar. We'll first set the title of the error window.

```
catch(Exception e)
        {
            new AlertDialog.Builder(this)
            .setTitle("Error")
        }
```

Next, we'll add an error message to the window to be displayed to the user.

```
catch(Exception e)
        {
            new AlertDialog.Builder(this)
            .setTitle("Error")
            .setMessage("There was a problem connecting to the Arduino")
        }
```

We'll now add two buttons to the error message, one allowing the user to accept the error, and one to attempt to reconnect to the Arduino.

```
catch(Exception e)
        {
            new AlertDialog.Builder(this)
            .setTitle("Error")
            .setMessage("There was a problem connecting to the Arduino")
            .setPositiveButton("Reconnect", new DialogInterface.OnClickListener() {
                public void onClick(DialogInterface dialog, int which) {
                    // Reconnection Code
                }
            })
```

```
        .setNegativeButton("Cancel", new DialogInterface.OnClickListener() {
            public void onClick(DialogInterface dialog, int which) {
                // Do Nothing
            }
        })
}
```

The `setPositiveButton` and `setNegativeButton` methods create new buttons that are attached to the error dialog. Inside the `onClick` methods, we can describe what will happen when the user clicks. We'll first call the `connect()` method again if the user asks to reconnect.

```
catch(Exception e)
    {
        new AlertDialog.Builder(this)
        .setTitle("Error")
        .setMessage("There was a problem connecting to the Arduino")
        .setPositiveButton("Reconnect", new DialogInterface.OnClickListener() {
            public void onClick(DialogInterface dialog, int which) {
                    connect();
            }
        })
        .setNegativeButton("Cancel", new DialogInterface.OnClickListener() {
            public void onClick(DialogInterface dialog, int which) {
                // do nothing
            }
        })

    }
```

If the user selects Cancel, we should close the app, so they will be able to relaunch and connect to their Arduino. Inside the `setNegativeButton`'s `onClick()` method, add the following line to tell Android that the program is finished executing.

```
System.exit(0);
```

Lastly, to show this alert to the user, append the `show();` method to the `AlertBuilder`. Note the semicolon to end the total series of methods. In total, your connect function should look like the snippet below.

```
public void connect()
    {
      try{
              soc = new Socket("192.168.1.60",23);
              in = soc.getInputStream();
              out = soc.getOutputStream();
      }catch(Exception e)
      {
          new AlertDialog.Builder(this)
          .setTitle("Error")
          .setMessage("There was a problem connecting to the Arduino")
          .setPositiveButton("Reconnect", new DialogInterface.OnClickListener() {
              public void onClick(DialogInterface dialog, int which) {
                      connect();
              }
          })
          .setNegativeButton("Cancel", new DialogInterface.OnClickListener() {
              public void onClick(DialogInterface dialog, int which) {
                  System.exit(0);                }
          })
          .show();
      }
    }
```

Now, if you launch the application (without an Arduino around), you should see the screen as shown in Figure 11.19. Clicking Reconnect will attempt to connect again, and clicking Cancel will close the application.

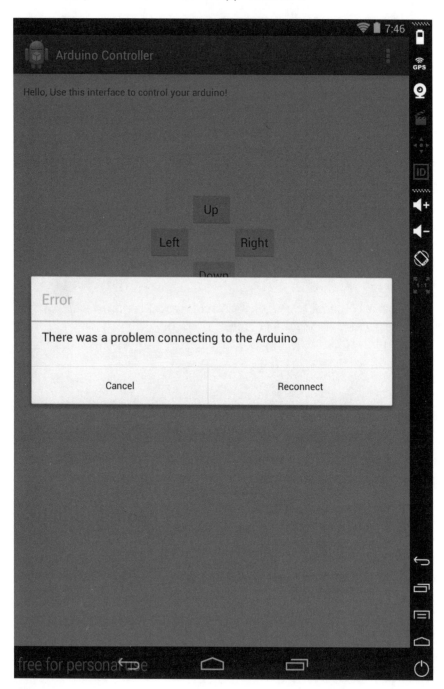

Figure 11.19
Arduino controller error dialog.
Source: Genymobile/Android

Now, we just need to provide functionality for each of the buttons on the primary screen. To do this, we'll need to get the buttons created by the View and attach `onClick()` methods to them. We'll do this in the `onCreate()` method, just below the call to our `connect()` method.

1. First, we'll need to get the button objects from the View. For the Up button, use this line of code:

```
final Button upButton = (Button) findViewById(R.id.upButton);
```

In this code snippet, we create a button variable called `upButton`. By adding final, we are saying that it will never change. The `findViewById()` method returns the button (or any other View element we request) based on the ID we pass it. These IDs are generated based on the names you give them in the Properties window, by the pre-compiler mentioned in the introduction. You will need to replace `upButton` with whatever you named your button. The system will attempt to help you by displaying all IDs containing whatever you type.

2. We'll now create the remaining button variables.

```
final Button upButton = (Button) findViewById(R.id.upButton);
final Button downButton = (Button) findViewById(R.id.downButton);
final Button leftButton = (Button) findViewById(R.id.leftButton);
final Button rightButton = (Button) findViewById(R.id.rightButton);
final Button stopButton = (Button) findViewById(R.id.stopButton);
```

3. For each button, we will need to assign a "listener" to listen for user action on each button. We'll do this by calling the `setOnClickListener()` method on each button.

```
upButton.setOnClickListener();
```

4. We must also pass a listener to the method to assign it. We'll therefore create a new listener. As soon as you create the new `View.OnClickListener()`, Eclipse will likely autofill the required `onClick()` method; if not, the correct signature is below.

```
upButton.setOnClickListener(new View.OnClickListener() {

                @Override
                public void onClick(View v) {
                        // TODO Auto-generated method stub

                }
        });
```

5. Inside the `onClick()` method, we'll fill in our logic to send a command to the Arduino. To do this, we'll use the OutputStream we gathered from our socket to write the forward command we previously programmed on the Arduino.

```
out.write('f');
```

6. Create these listeners for each button using the appropriate op-codes that we created in Chapter 8. Table 11.1 contains these codes.

Table 11.1   Op-Codes for Arduino Server

Op-Code	Function
f	Move Forward
b	Move Backward
l	Turn Left
r	Turn Right
s	Stop

A completed `onCreate()` method is shown below.

```
protected void onCreate(Bundle savedInstanceState) {
        super.onCreate(savedInstanceState);
        setContentView(R.layout.activity_main);
        connect();
        final Button upButton = (Button) findViewById(R.id.upButton);
        final Button downButton = (Button) findViewById(R.id.downButton);
        final Button leftButton = (Button) findViewById(R.id.leftButton);
        final Button rightButton = (Button) findViewById(R.id.rightButton);
        final Button stopButton = (Button) findViewById(R.id.stopButton);
        upButton.setOnClickListener(new View.OnClickListener() {

            @Override
            public void onClick(View v) {

                out.write('f');

            }
```

```
                   });
                   downButton.setOnClickListener(new View.OnClickListener() {

                           @Override
                           public void onClick(View v) {

                                   out.write('b');

                           }
                   });
                   leftButton.setOnClickListener(new View.OnClickListener() {

                           @Override
                           public void onClick(View v) {

                                   out.write('l');

                           }
                   });
                   rightButton.setOnClickListener(new View.OnClickListener() {

                           @Override
                           public void onClick(View v) {

                                   out.write('r');

                           }
                   });
                   stopButton.setOnClickListener(new View.OnClickListener() {

                           @Override
                           public void onClick(View v) {

                                   out.write('s');

                           }
                   });
           }
```

Your app is now ready to connect to your Arduino. Ensure that your Arduino is loaded with the Wi-Fi sketch from Chapter 8 and turn it on. Check your router or connect from your Chapter 8 application to confirm that the IP address associated with your Arduino has not changed. If it has, update the IP address in the MainActivity.java file's connect() method. Launch the app in either the emulator or your physical phone and remotely control your Arduino!

Conclusion

In this chapter, you successfully installed all the components of an Arduino development environment, created an Android app to connect to your Arduino, and ran it on both the simulator and a real-world phone. These skills can not only be used to extend Arduino capabilities, but they provide a cornerstone for the development of all other Arduino and mobile apps as well.

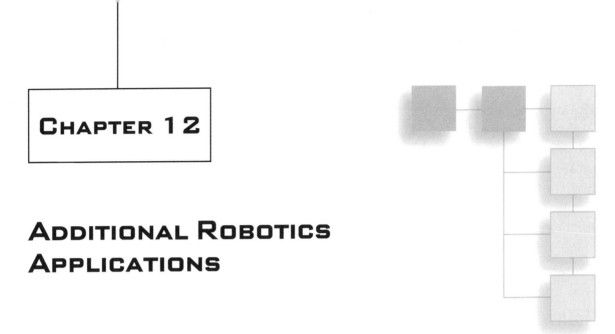

CHAPTER 12

ADDITIONAL ROBOTICS APPLICATIONS

Robotics as a discipline has made tremendous progress and permeates many facets of society. Robots are used in medicine, education, the military, commercial applications, and many other fields. This chapter highlights some of the strides being made in the applications and use of robotics in these fields. Additionally, some ideas on extending projects discussed in the book are also presented.

ROBOTS IN MEDICINE

Robots have come a long way in medical applications. Robotic surgery is being used in many hospitals and research institutions. Robots are being used in neurosurgery, orthopedic surgery, laparoscopy, and other areas. For more details about these projects, visit www.hindawi.com/journals/jr/2012/401613/.

Tele-robotics is another application of robotics in medicine. In tele-robotic applications, the surgeon controls the robot from an external user interface, and the robot performs the surgery. In the Robotic Surgery Center of NYU Langone Medical Center (http://robotic-surgery.med.nyu.edu/for-patients/what-robotic-surgery) surgeons use the "da Vinci Si" robot equipped with four arms to perform surgeries with miniaturized instruments that can be controlled via a user interface. Their goal is to achieve minimal invasion into a human body, thus leading to less trauma, minimal scarring, and faster recovery time.

After reading the previous chapters of this book, it is not hard to imagine a robot performing surgery, since you have seen how to equip a robot with various sensors, tools, and cameras, and you also saw how they can be controlled via Bluetooth or Wi-Fi. Of course, building such robots is by no means an easy task.

However, as an extension to the medical application project you completed in this book, you can easily imagine building a robot that works as a nurse's assistant. You can imagine the robot performing all the tasks a nurse does during a patient's visit to the doctor's office, such as recording vitals (e.g., pulse, blood pressure, temperature, and weight and height of the patient) and then forwarding them to the doctor. It will take some time to complete such a project, but the introduction provided in the chapters of this book should establish a good start for you to work on this project.

ROBOTS IN EDUCATION

Robots have been used in education for a long time. The two most popular companies that promote robotics for educational purposes in K-12 are Lego and Vex robotics. Other venues for exposing K-12 students to robotics are after-school activities and summer camps. Through such exposure, students develop an interest in science, so these activities are promoted for that purpose.

Some of the exciting projects that students in K-12 are working on around the country include building rockets and underwater submarines. Students in Archbishop Murphy High School in Seattle, Washington (www.washingtontimes.com/news/2014/may/24/students-learn-physics-by-building-rockets-robots/?page=all) have built underwater submarines as part of their physics class and have even tested these robots.

In higher education, robotics is taught as part of computer science-, engineering-, and technology-related courses. Additionally, competitions at regional, national, and international levels are held to further encourage students' interest in robotics. Some of the research outcomes of such courses and competitions have contributed to novel designs and applications of robotics.

Recognizing the need for electrical and computer engineering technologists as part of the growing workforce some states, such as Michigan, are building a robotics program in collaboration with the industry so that students have the skill sets required to join the workforce when they graduate from college.

Some of the projects that resulted from research in educational institutions include:

■ Robotlabs. This company produces a variety of robots that can be used for teaching science and engineering topics. They also produce a humanoid robot and a quadcopter robot that are used to teach math and science concepts.

■ Muscle tissue for robots. Researchers have developed artificial muscle that can lift weights up to 80 times more than its own weight. (www.gizmag.com/artificial-muscles-robots-nus/28923/). The artificial muscle was also used in designing fish that can swim underwater. The fabrication and development of muscle tissue represents a major achievement in the area of robotics.

ROBOTS IN THE MILITARY AND LAW ENFORCEMENT

Robots are increasingly being used in situations that are hazardous and harmful to humans in military reconnaissance scenarios. Several types of robots are being used for various purposes. There are small simple robots that can be thrown into potentially dangerous situations—for example, to diffuse a bomb in an arena or to search for chemical weapons, and so on.

Robots used for such purposes are equipped with sophisticated sensors, cameras, Wi-Fi cards, and other tools to provide the required information to officials in charge of the operations. Reliability and security of the information gathered by the robot is also extremely important in these mission-critical situations, and therefore plenty of research and testing is done prior to implementation.

A classic example of using a robot in a law enforcement situation is exemplified in the most recent Boston bombing incident. In this case, the police sent an industrial robot into the boat where bombing suspect Dzhokhar Tsarnaev was hiding. The authorities wanted to make sure that there were no explosives strapped to him that could explode at any time. The robot was capable of picking up any such explosives to clear the way for the police authorities. Once the authorities received information from the robot that everything was safe, they moved in to capture Tsarnaev.

Other military robots include aerial unmanned predators and drones. These robots can extend their limbs and utilize sight and other tools with which they are equipped to perform various tasks and take out the enemy. They can go into potentially dangerous situations and alert soldiers on the conditions inside a battle area before soldiers actually get there.

The security robot project in this book presents the idea of attaching cameras to robots that can take and store photographs. From here, you can imagine how similar robots can scan QR codes and other images that can be processed by additional software or mathematical algorithms and used by military and law enforcement agencies.

Robots in Industrial Applications

Industrial robots are also becoming very popular. The market for industrial robots is expected to reach $41.17 billion by 2020 according to a report by Allied Market Research (http://roboticstomorrow.com/article/2014/06/industrial-robotics-market-is-expected-to-reach-4117-billion-globally-by-2020/266).

Reports from *Science Daily* on the progress made in this field (www.sciencedaily.com/articles/i/industrial_robot.htm) suggest that several types of robots are being manufactured in a variety of shapes and sizes for a variety of purposes. Robots are shaped as animals, insects, humans, or other types for use in situations where that particular shape works best. For example, a spider-shaped robot mimics the behavior of that arthropod and is used to weave and navigate across a "web," which another species of robot may be unable to accomplish. And since these are machines, you can actually enhance the capabilities of that species by adding extra features through sensors and peripheral devices. Some of the purposes for which industrial robots are used include:

- Restocking agents in retail stores like Home Depot, Lowes, and Amazon
- Surgeons and medical assistants in health care
- Law enforcement assistants
- Automotive assembly and manufacturing
- Household activities (vacuum cleaners, home automation, personal assistants, etc.)

Trends in Robot Types

Recently, there has been some remarkable advancement and activity in robot design. Until recently, most robots were built using hard casings and were therefore quite rigid. However, robotics engineers at various locations, including MIT, have now used soft and flexible materials in designing robots, thus providing scope for new applications of robotics. Engineers have also further developed the coordination of multiple robots, referred to as swarm robots. Some of these trends are discussed below.

Soft Robotics

As robots permeate our daily lives, scientists are looking toward robots that are more human friendly by being soft, agile, and maneuverable, and that can tolerate errors when landing on surfaces. Soft robotics has become a recent research activity where researchers are designing robots that have soft bodies. Some even have fluids flowing through their bodies to provide all the features mentioned above. Among their many advantages, these soft robots don't cause harm to other objects when they bump into them. Similarly, because they are flexible they can mold themselves to fit into tight spaces.

Some example projects of soft robotics include:

- A robot fish designed by MIT that moves like a fish and rotates with almost the same speed of a real fish (https://newsoffice.mit.edu/2014/soft-robotic-fish-moves-like-the-real-thing-0313)

- A robot that can grip an object through efficient information processing by eliminating feedback required from the main program (http://jfi.uchicago.edu/~jaeger/group/Soft_Robotics/Soft_Robotics.html)

Swarm Robots

Swarm robotics is a new area of research that studies the coordination of multiple simple robots. The difference between swarm algorithms and traditional applications of multiple robots is that swarms by nature are not centralized—meaning there is no master robot coordinating all the others. Swarms accomplish a task by sheer number of attempts and a feedback loop that alerts other agents when a task is complete or when a task requires more workers. Many of the algorithms behind swarming applications are gained from the natural world, such as ants, which are able to accomplish tasks without direct communication, through the buildup of pheromones. As more and more ants discover an attacker, more and more "attack" pheromone is released, and increasingly more ants move to defend.

Researchers in robotics are attempting to study this group behavior and factors that lead to a group's success and apply it to robots. One area of study is focused on allowing robots to self assemble into groups based on the needs and tasks that arise and acclimate to the situation. Researchers at MIT (www.kurzweilai.net/mit-inventor-unleashes-hundreds-of-self-assembling-cube-swarmbots) have created robots that have no external moving parts but can climb over one another, jump in the air, and perform several other acrobatic feats. The goal of this research is to design robots that can take any form or shape, and have a bunch of them swarm together and tackle emergencies. These ideas may seem futuristic, but current research is demonstrating the initial development in this area.

CONCLUSION

As discussed in this chapter, the difficulty in building these robots for various applications is that they are very complex projects. The design of such robots requires specialists from several fields, such as biology, mechanical engineering, physics, material science, and other professional disciplines. However, the projects discussed in this book should give you a good introduction and hopefully will motivate you to further explore robotics and perhaps build the next robot that can change the world.

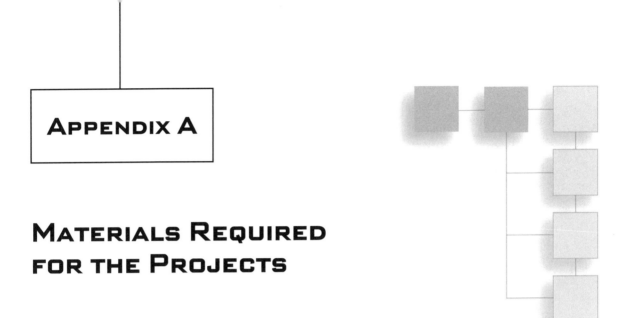

Appendix A

Materials Required for the Projects

Chapter 2

- Arduino Uno R3 board (Amazon, SparkFun, Adafruit)
- Magician Chassis kit (Amazon, SparkFun)
- Ardumoto board (Amazon, SparkFun)
- Ultrasonic range sensor model HC-SR04 (Amazon)
- 9-volt battery
- Jumper wires with connectors
- Solderless breadboard, plug-in type (Amazon)
- Piece of dust cloth (e.g., Swiffer)

Chapter 3

- QRE1113 Analog line sensors (2)
- Robot chassis with Arduino board and Ardumoto motor control shield (as built and used in Chapter 2)
- TCS34725 color sensor

CHAPTER 4

- Arduino board
- Laser diode sensor
- Photoresistor sensor
- Ultrasonic range sensor
- LEDs of different colors
- Solderless breadboard and hookup wires

CHAPTER 5

- Standard Arduino robot used in previous chapters
- Arduino Wi-Fi shield
- USB mini cable
- Ultrasonic range finder
- Wi-Fi network

CHAPTER 6

- Arduino board
- LCD display
- Push button
- Trimmer potentiometers (2)
- Speaker
- Solderless breadboard and hookup wires

CHAPTER 7

- TMP36 temperature sensor
- 16×2 LCD panel + 10 kΩ potentiometer
- Arduino board and solderless breadboard
- Stackable SD card reader
- SD card (not SDHC card)

Chapter 8

- Standard Arduino robot used in previous chapters
- Arduino Wi-Fi shield
- USB mini cable
- Ultrasonic range finder
- Wi-Fi network
- Computer with Internet access
- NetBeans with JDK (free software)

Chapter 9

- Standard Arduino robot used in previous chapters
- RadioShack camera shield
- SD card shield with SD card
- Computer with SD card reader

Chapter 10

- Arduino board
- NeoPixel ring. A 12-LED ring from Adafruit.com: 12 x WS2812 5050 RGB LED ring.
- Sound sensor (microphone). One of the three sensors in the Robotics Sensor Kit from RadioShack. The kit also contains an optical sensor and an IR sensor.
- Solderless breadboard and hookup wires

Chapter 11

- Standard Arduino robot used in previous chapters
- Arduino Wi-Fi shield
- USB mini cable
- Android phone (or free Android emulator)
- Wi-Fi network
- Computer with Internet access
- Android Developer Kit (free)

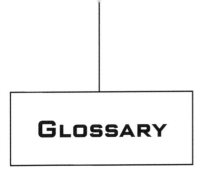

GLOSSARY

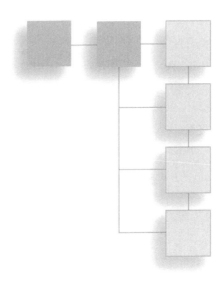

AJAX Asynchronous JavaScript and XML

algorithm An algorithm is written to perform calculations in an ordered fashion to achieve an objective.

ambient temperature refers to the temperature outside

Arduino an open-source electronics platform

Ardumoto a motor shield that can control two DC motors

Artificial Intelligence (AI) Machines with intelligence built in with software show artificial intelligence.

Baud rate speed at which data is transferred

Bluetooth This technology uses short wavelength radio waves for establishing wireless connections over short distances.

CCD Charge-Coupled Device

CMOS Complementary Metal-Oxide-Semiconductor

color sensor Color sensors give varying outputs depending on the colors to which they are exposed.

CSS Cascading Style Sheets

drones unmanned aerial vehicles with extensive military applications

DSLR Digital Single-Lens Reflex (camera)

FXML JavaFX Extensible Markup Language

GUI Graphical User Interface

HTML HyperText Markup Language is the standard markup language used to create web pages.

humanoid A humanoid robot is a robot with its body shape built to resemble that of the human body.

IDE Integrated Development Environment; a tool for software development

IEEE 802.11 802.11 is the generic name of a family of standards for wireless networking related to Wi-Fi.

IP address Internet Protocol address is the label assigned to each device in a computer network.

IR emitter detector an infrared sensor that sends out an infrared signal and receives the reflected energy from an object

Java Java is a programming environment that allows you to play online games, chat with people around the world, calculate your mortgage interest, and view images in 3D, just to name a few examples.

JavaFX variant of Java, extensively used in RIA (Rich Internet Applications)

JavaScript JavaScript is the programming language of the Web.

JDK Java Development Kit

LAN Local Area Network is a computer network that interconnects computers within a limited area like a home or school.

LCD Liquid Crystal Display

LED Light-Emitting Diode

MAC address The Media Access Control address is a unique value associated with a network adapter composed of 12-digit hexadecimal numbers.

Magician Chassis A kit manufactured by SparkFun for use with Arduino boards. The chassis comes with a set of two wheels and a platform for mounting the Arduino board and associated shields.

MATLAB MATLAB is a high-level technical computing language and interactive environment for algorithm development.

modem a modulator/demodulator used to encrypt analog data for transmitting over a wireless network

NetBeans IDE NetBeans is an Integrated Development Environment (IDE) for development primarily with Java. It is also used with other languages such as PHP and C/C++.

Ni-MH batteries nickel-metal hydride rechargeable batteries

pairing establishing a connection between two Bluetooth devices

pheromones Pheromones are chemicals released by an organism into its environment, enabling it to communicate with other members of its own species.

photoresistor a sensor whose resistance changes with changes in the amount of light falling on it

PHP PHP is an acronym for "PHP Hypertext Preprocessor"; PHP is a widely used, open-source scripting language.

piezo-buzzer a device that sends a beep/makes noise

proximity alarm a device which produces an alarm when someone is in the vicinity

Python a programming language

Qt Qt is a cross-platform application and UI framework for developers using C++ or QML, a CSS & JavaScript-like language.

quadcopter a helicopter that is lifted and propelled by four rotors

router A networking device that forwards data packets between computer networks

SD card Secure Digital non-volatile memory card

shield commonly used to denote an add-on board in Arduino hardware

SparkFun a company that manufactures many products used in projects with the Arduino boards

Swing Swing is the primary Java GUI widget toolkit.

syntax In programming, syntax refers to the rules that specify the correct combined sequence of symbols that can be used to form a correctly structured program using a given programming language. Programmers communicate with computers through the correctly structured syntax, semantics, and grammar of a programming language.

telemedicine the practice of medicine using wireless techniques

telerobotics the area of robotics concerned with the control of semi-autonomous robots from a distance, chiefly using wireless networks

Telnet a network protocol that provides a bidirectional interactive text-oriented communication facility

TO-92 package a 3-pin plastic header-style package used mainly for transistors

torque Torque is a measure of how much a force acting on an object causes that object to rotate.

touch sensor a sensor that gives out a signal when someone touches it

UART Universal Asynchronous Receiver/Transmitter

ultrasonic sensor A sensor that emits an ultrasonic signal and detects the reflected signal. The time between the signal being sent and being received gives a measure of the distance of the object.

untethered connection In untethered communications, there are no physical connections using wires and, as such, there are no constraints.

USB Universal Serial Bus

Velcro fabric fasteners

Wi-Fi local area wireless technology that enables an electronic device to connect to the Internet

INDEX

Look for complete video tutorials for each project in this book on www.cengageptr.com.